The Hypothesis on the Edge of Science

ABOUT THE AUTHOR

Dr Jewsbury is a retired theoretical physicist. He was born and educated in England. However, he has lived in Australia since 1977. He has been a professional scientist for over thirty years. He was employed by universities in England and Australia for a decade. Then more recently he worked in multi-disciplinary scientific research and development for over two decades. He is the author of many technical scientific papers. This is his second book. In this book, he presents a discussion on the nature of the ultimate reality. These ideas are a distillation of his accumulated thoughts and reflections over many years. The book describes how a new hypothesis, sitting on the edge of science, can explain the existence of all things including intelligent life, science, logic and the universe.

The Hypothesis on the Edge of Science

An Explanation of Eventual Reality

Peter Jewsbury

Email: peter.jewsbury@gmail.com

2008

Published by *A*libro *A*ustralis *S*cientifica, Melbourne, Australia 2008.

First published in Australia in 2008 by *A*libro *A*ustralis *S*cientifica.

Description: x, 166 p. 8 ill., 5 tbl. ; 24 cm.

Cataloguing:
Fields: Foundations of knowledge, philosophy of science, ontology.

Search keywords: fundamental, reality, eventual reality, mind-body, logic, existence, langue, ultimate question, metaphysics.

National Library of Australia Cataloguing-in-Publication entry:

Author: Jewsbury, Peter.

Title: The hypothesis on the edge of science : an explanation of eventual reality / Peter Jewsbury.

Publisher: Essendon, Vic. : *A*libro *A*ustralis *S*cientifica, 2008.

ISBN: 9780980315127 (pbk.)

Notes: Includes index.

Subjects: Science--Philosophy.

Dewey Number: 501

CONTENTS

List of illustrations

List of Tables

"Apeiron's Dawn"

Where e'er the light of dawn doth spread
'cross a morass of mental slime,
A spark of joy, of puzzled dread,
Speaks of splintered paradigm

Demands to know of what is read,
'Till all at once the answer is,
So undoubtable, yet clearly said,
The 'peiron of bland choice, and bliss.

And still when in some pensive mood,
I dwell upon the picture shown.
How can the answer be so crude,
Yet not be seen as overblown?

Lo, all mysteries are thrown out,
Which have evaded all before.
Not an ounce of forgotten doubt
Remains to sour the bon humour.

Thus the edifice doth beguile,
Without support of any style.
Hidden answers to nature, smile
For all to see and reconcile.

PREFACE

It is in our basic nature to be curious, to ask profound questions and to seek their answers. However, sometimes we can have great difficulty framing a question appropriately. Thus asking the right question does indeed take you half way to the solution.

Like many of my readers, I have spent half a lifespan passively wondering about the great questions of existence – simply awaiting inspiration. Certainly like Pasteur, one supposes that only the prepared mind can ever hope of even glimpsing a partial answer. Very occasionally I thought actively on the topic. I have a charming collection of naïve writings from my early teens. In my thirties, I twice gave a reasonably conventional undergraduate course on the topic. Then in my fifties, I returned to complete the work with a clearer understanding of the question, renewed enthusiasm, knowledge and conviction.

It seems that some problems have been handed down unsolved through the generations from antiquity. These problems can now appear so intractable that no sane person would expect that they could answer them. Inevitably such despondent thoughts sap initiative. This attitude destroys progress. Lack of progress breeds resentment and criticism of more successful activities. Thus one put-down of science claims that science is successful just because science seeks only to answer the answerable questions. Of course this statement is true. However, we should look at the reason why it is true. Apparently every properly framed question turns out to be answerable. Experience teaches that we generally only succeed in answering half the problem. Then the other half of the problem gives rise to a new question. This is why doing scientific research is such a fascinating game. Indeed many of the great scientists have remarked how amazing it is that the world is so understandable. Such a confident attitude bolsters efforts. These carefree efforts always seem to succeed despite the opinions of pessimists.

So what is the hypothesis in this book? It is a new rational self-evident explanation of everything. Is it fanciful speculation? Is it opinion based upon facts? Well, it is not opinion based upon facts. The hypothesis relies upon no facts. Likewise the hypothesis is not fanciful speculation. All results are derived and justified. Furthermore it is not hard science either. Certainly the hypothesis does give rise to known laws of science. However, what you will find in this book is just a carefully reasoned argument. The argument reveals some key fundamental characteristics of *the* great question and gives an ultimate solution. Naturally, in some sense, this book only provides an introduction to a much longer answer. This manuscript describes the structure, mechanisms and basis of reality. Clearly there is much more work that needs to be explained. Nevertheless the restricted answer given here is compelling.

This book is intended as a simple introduction to the hypothesis for the intelligent general reader. Hence the presentation is given with almost no mathematics. A highly mathematical treatment of the physical aspects has been given elsewhere[1]. Naturally 'simple' is a relative word. You should expect to be bombarded with a bevy of beautifully strange ideas. Hence please do not rush the reading of important sections. There is fun to be had in the journey, savouring an anticipation of the answer. Furthermore do not be disappointed at the inevitable simplicity of the conclusions. The solution really has to be simple.

You will note that a few sections of this book have a grey background. Greying designates a harder section that readers may safely ignore. Only advanced readers, physical scientists, philosophers and mathematicians are encouraged to read these sections.

Melbourne Feb 08

---

1 Jewsbury P *Resolution of Epistemological Analytical and Logical Issues in the Theories of Yesterday* (*A*libro *A*ustralis *S*cientifica 2006) 280pp.

PROLOGUE

Have you ever stood on firm ground, looked up at the sky, felt the sun's rays on your skin with a breeze ruffling your hair, smelt the scent of pollen in the air and wondered whether the perceived world is simply a façade. Unless you are a philosopher or a storyteller the chances are you have not. Why should you? It is so easy to embrace the mindset of conventional wisdom, a mindset built up over millennia of careful study by some of the greatest minds of recorded history. Surely it is obvious that reality is just space and matter, energy and motion, and action and causal reaction. Is not it? Well, what are the other possible viewpoints?

We could choose from many ridiculously implausible alternative schemes. However, I am not going to hypothesize that we live in a computer simulation, a 'Matrix' of science fantasy; neither will I claim that our whole world exists only in the mind of a living deity of inconceivable power. I will not argue that this world of matter, energy and space does not exist. Of course it makes no sense to deny the existence of an exterior world of abundant forms. This magnificent external world is the world that challenges us. Yet there is an important question buried beneath all the idle speculation. Is this observable stuff of the universe, all that is? Is current science seeing everything that actually exists or just a superficial shell standing between an ultimate reality and our extended senses. Furthermore if there is something more to reality than meets the eye, then could current science be missing almost everything that exists?

We shall be arguing decisively that the answer is yes. We will finally expose this deeper reality. It seems that for centuries we have been refusing to accept the existence of this obscured reality. Part of the reason for this wanton denial lies in the unfamiliar appearance of some aspects of the new ideas. After all, we are creatures of habit and sometimes completely miss what lies beneath our noses. Indeed we have behaved like this many times before, ignoring the inevitable truths. Thus let us refer to our history to see why some ideas can initially be overlooked as too disturbing.

There was a time when it was comforting to think that the ground under our feet never moved (apart from the rare jolt). We thought that the Earth sat at the centre of a universe revolving around us. We were told that the universe was made for mankind. Nowadays we have come to accept that this Earth-centric and human-centric view is not true. What's more, when we came to accept the truth, the sky did not fall in. In fact, most of the time, we still conveniently reference ourselves with respect to Earthly features just as though the Earth were immobile. Thus today, the movement of the Earth is neither an intriguing mystery nor a contentious problem.

Part of growing up as a scientifically literate species (*animans scientifica*) is coming to terms with such apparently unappealing facts. We have created our science but not our universe. Our understanding of science does make us special but we are not the crowned rulers of creation. It is almost a certainty that there are innumerable extraterrestrial alien civilizations whose science far surpasses ours. Furthermore no working science is fixed – science relentlessly grows in knowledge and understanding. In particular, science gives us a series of convenient representations, which we expand and, from time to time, dramatically improve. We should expect that, at its first introduction, each more basic description would probably seem uncomfortable too. However, we need to remember that (by expanding our appreciation of the world) nothing serious will ever have actually altered. Generally nature will still obey the older familiar set of scientific laws (in appropriate contexts). Nevertheless, with each advance, we will have understood our place and role in the universe a little better.

Thus history records that new fundamental ideas often seem contentious and disturbing. For instance the classical scientific view (ie the view that was founded by Sir Isaac Newton) apparently ruled out free will. This is an old worry, which has never really been resolved. More recently in the twentieth century, quantum mechanics appears to remove support for true certainty in our world. Let me illustrate this undermining of a definite reality by recalling the fable of Schrödinger's cat. We imagine placing a cat in an impenetrable box with a phial of cyanide and a trigger device. The trigger device responds to the unpredictable decay of a radioactive nucleus. The radioactive decay activates the trigger to release the cyanide and the cat dies supposedly immediately. Now, we turn on this hypothetical device by sealing the box. Then we wait.

Quantum mechanics tells us that the cat, inside the sealed box, is neither alive nor dead but in an in-between state ripe with both possibilities. This indecision is the complete description of the quantum state; no as-yet-unknown (local) variable may actually record the fate of this unfortunate cat before we open the impenetrable box. Experiments on electrons, photons and atoms (but not cats) have confirmed this behaviour. In this discussion, I am ignoring the question as to when a measurement occurs. Actually the cat should be able to make her own measurement. However, I want to keep everything at the quantum level. Thus we should imagine either using a quantum cat and trigger device or simply replace the cat by an atom which can be disintegrated by the radioactive decay. Anyway, why nature behaves in this startlingly uncertain manner is the uncomfortable mystery. This mystery has remained for the successor of quantum mechanics to answer.

Clearly there must be a successor. Science is a serial work like a wondrous epic saga spread over many mystery novels. Individual novels leave loose ends that later novels explain. Only the last novel solves all the remaining puzzles and gives the definitive answers to obscure mysteries underpinning the whole saga. In this simile, the final volume of science's story describes the hypothesis of eventual reality.

Eventual reality explains all those unsavoury features that were left unresolved by earlier volumes. Thus unusually, eventual reality is more inspiring than disturbing. The hypothesis easily ties up the just-mentioned loose ends. Hence we find that the universe not only allows free will, the universe demands free will. Indeed the universe is an expression of unbridled opportunity. Furthermore Schrödinger's (quantum) cat is in an ill-determined state because (anti-causally) future events will help to determine her current perilous medical status. These are some of the confronting conclusions of eventual reality but we need to start our explanations at the beginning.

Thus in this book we explain why there must be more to the universe than is shown in the current scientific picture of reality. I will take you on an intriguing journey venturing beyond the bounds of science to a hypothesis sitting on the edge of science, the hypothesis of eventual reality. Inevitably I may have attempted to explain rather too much in too few pages. So do not despair if you find some parts difficult to follow. Many readers will find it useful to reread some of the later chapters several times before proceeding to the next chapter. Nevertheless I hope that you will still find the book exhilarating.

Now, the book contains a general discussion of a restricted range of quite fascinating fundamental topics. For instance I explain why the universe must exist and what the universe's ultimate form looks like. I discuss why the world is rational and understandable, and how nature is transformed into the observed forms. Note that an earlier book gives many-more technical details, which will most likely only be fully accessible to a theoretical physicist. As time has rolled by, my thoughts have become clearer and thus I am able to write this much simpler explanation introducing the hypothesis.

Of course, since the birth of science, finding an answer to the question of ultimate reality has mostly been a pipe dream – a hope well beyond reach. Only careless thinkers, misguided cranks, religious zealots, attention seekers and conmen have pedalled their definitive versions of the ways of the world. Thus you might think that a final answer is an impossible dream like the elixir of life, the perpetual-motion machine and the alchemist's sought-after base-metal transmutations induced by a mystical Philosopher's stone. Well, actually there is a critical difference between these extreme dreams and the search for an ultimate reality. The only reason why we have tended to ignore looking for an ultimate answer is that we do not think we can find one, and not that an answer does not exist. In fact an answer does exist, as we shall see.

Now, sometimes abundant choice can spoil us. There is an old saying that all roads lead to Rome. Of course that old set of roads (radiating across Europe from Rome) were built after the city of Rome was founded. However, there must have been an even older and perhaps less formal route that took the founders to the city's site. Similarly there are now several formal approaches (and a founding approach) that reveals the fundamental details of this hypothesis. Here we shall present one of these formal approaches, only obvious in retrospect. Nevertheless like all major roads this route is designed to allow an easy passage for the typical journeyer and fails to acknowledge properly the obstacles faced by the earliest travellers.

Hence let us turn our attention to the organization of this document. Do not be misled into thinking that I am going to present what many like to call a scientific theory of everything. Scientific theories, sitting within the material paradigm, cannot explain themselves or the reason for existence. No. I shall describe a hypothesis sitting on the edge of science, a hypothesis that can explain itself, a hypothesis that can explain the reason for existence and incidentally a hypothesis that can explain all of science.

The first two chapters provide an introduction. We start our telling of the story with the magnificent vista formed from the collective wisdom of glorious ages. It is breathtaking to view the full historical picture of scientific ideas stretching back over the last two and half millennia. Thus in chapter one, we survey a selection of the interesting and important, ideas and suggestions appearing in the surviving recorded history of ideas.

Now, we use many words all the time without any full appreciation of what the words really mean. Thus chapter two focuses on the meaning of a few critical terms. We introduce definitions for the interrelated concepts of reality, science and the universe. Later chapters build on these definitions to give further clear fundamental explanations of concepts like space, matter and mind. Essentially these first two chapters lead you across familiar ground to a precipice that marks the boundary of conventional reality.

Then in the next three chapters I ask that you open your mind to a new exciting view of reality. Let reason and logic be your guide, not obsolete belief. On casual reflection it almost seems non-sensible to think that we could rationally explain the existence of the universe and everything the universe holds even including the nature and applicability of logic. Yet we reveal one methodology that can accomplish this aim. Thus the third chapter describes how this special technique works. However, the complete proof of absolutely everything's existence takes a further two chapters to complete. Nevertheless we are able almost immediately to see what the fundamental universe must look like.

This fundamental universe, the apeiron, is non-spatial. However, in chapter four we define space and show how we discover space as an attribute of the fundamental universe. This foundation space has one companion. Thus a mysterious aether, holding untold secrets, fills this foundation space.

Then, we are able to complete the *by inspection* proof. Intriguingly we find that two logical tautologies bootstrap rationality and logic into existence. This procedure delivers the form of natural logic and it completes the central proof. Thus in chapter five, we finalize the justification for the existence of everything including logic, space and the universe.

From this point in the book we begin to see more recognizable aspects of the scientific world. We discuss some empirical data from the scientific world. Hence, in chapter six, we define matter and explain several of its features. We also explore the empirical relation between thought and mind. This helps to clarify the intended meanings of mental terms that we are using.

Now, there are no prior scientifically credible fundamental explanations of the mind. However, by peering into the picture derived from the current hypothesis, we can identify clearly what constitutes a brain and a mind. Thus in chapter seven, we reveal the intrinsic vocabulary, grammar and logic of the langue of the universe. We explain free will and its significance for consciousness. Hence we discover the real reason why there is a capacity for intelligence in the universe – the exploitation of opportunity. Matter adds continuity and certainty, and mind provides for opportunist design and innovation.

Then, in chapter eight, we end this book with a cautious assessment about the place of science in today's society. There are certain historical parallels between societal circumstances today and happenings in the distant past. We must be on guard against complacency. There is no certainty that science will always be available. It seems that science is losing prestige. Yet at the same time, advanced societies are completely dependent on the health of science. Hence we must properly value science. Otherwise we face the prospect of losing access to essential scientific knowledge, learning and wisdom. Indeed politicians must adapt to listen and respond appropriately to the messages of science. In turn, scientists must put more effort into public education. In the modern world the alternative is to wait for evolution to find a fitter life form to rule planet Earth and her surrounds. The prime lesson is clear, better educate everyone in science or perish.

1. Every Story has a Beginning

In this book I will try to impart to you an understandable answer to an utterly amazing question. What is the ultimate explanation for everything? You will be aware that this is not a new question. The curious inquisitors have rephrased and part phrased this quandary in many ways: What is the meaning of, the purpose of and the reason for the existence of everything? Why does the universe exist? Could any feature of the universe have been any different from the way it is? Why is there life in the universe? Why is the universe understandable?

Thus over millennia, the curious have proffered many crude answers for this ultimate question. Indeed surviving written records, formed since the beginning of technical writing, show surprisingly useful deep insights into the fundamental universe. Undoubtedly many other insights have been lost from the records; they have been either dismissed or destroyed and forgotten. However, many snippets of old revelations remain to intrigue and astonish us. It is remarkable that these ancient sages knew or suspected so much useful knowledge so long ago.

Hence let us make a quick survey of some of the highlights of many astonishing ideas from bygone ages. These ideas are dazzling gems, veritable pearls of wisdom. Figuratively this summary describes a gleaming pearl necklace stretched out over time and laid across an unimaginably larger starry field. The highly restricted survey will allow us to understand specially selected aspects of our intellectual heritage. It will allow us to appreciate why the answer to the fundamental question can be so simple; yet, at first glance, this answer may appear far from obvious.

The Greeks

We shall start our deliberations 2600 years ago along the sunny sea-lapped shores of the Eastern Mediterranean. Over the several centuries that followed, notable Greek philosophers of antiquity devised (often-conflicting) schematic pictures of reality.

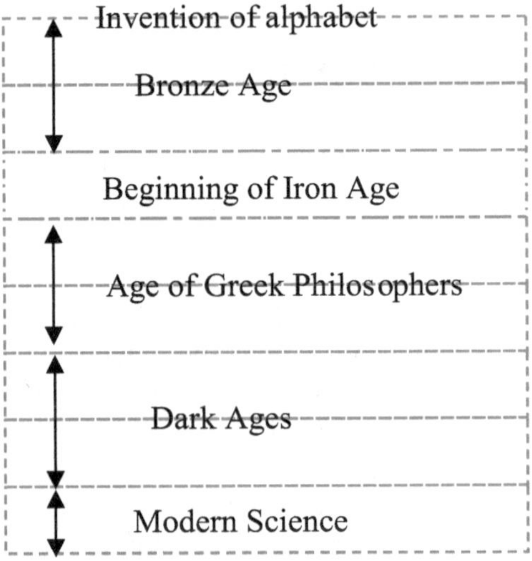

Figure 1: This figure roughly depicts those ages of modern civilization relevant to our story. The scale is marked in approximate five-hundred year intervals.

Why start this discourse so long ago in a place I have never visited? Quite simply, ancient Greece remains subliminally critical to any understanding of so-called Western civilization. Respect ensues from any appeal to an ancient-Greek connection. Scientists will recall that their colleagues enlisted Greek views on science in the debates about Dalton's atomic hypothesis and the meaning of quantum mechanics. Historians will recall how European empires drew inspiration from Greece. It is an easy task to entwine seamlessly mythology, propaganda and fact. In one story telling, Trojan Prince Aeneas (cousin of Hector) founded the city of Rome and indirectly her empire. In another telling, the Trojan Prince and Roman Consul, Brutus (grandson of Prince Aeneas), founded the nation of Britain and indirectly her empire. Linguists know how much the English language owes to words of Greek origin. Greek words and values are for ever present awaiting the opportunity for a speaker to utter them. Thus in the West, the influence is palpable. We still learn about Aesop's fables and the names of Greek deities. We could learn about many other rich cultural heritages but it is towards the Greeks that we turn. Let me remind you of this heritage by mentioning: the face that launched a thousand ships, Achilles' heel, Greeks bearing gifts, the Golden Fleece, the Olympics, Zeus, Hades, Hercules, the Elysian Fields, the Oracle at Delphi, Jason and the Argonauts, and Homer's Iliad as examples. This list only shows that there is a modern impact but does not explain why. However, the reason is obvious. We imbibe ancient Greek culture because some events of two and half millennia ago still matter to us today.

What were those events and why are they still significant? Well some things are inevitable but details of where and when these things will happen are quite uncertain. Very few individuals have the privilege of steering the course of human history. I am not thinking of the transient effects of dominant leaders and their military victories. I am thinking of rare individuals who irreversibly change everything for everyone. Thales (624-546 BC) was one of those individuals.

Let me take you back to those halcyon days on the shores of the Aegean. It was a peaceful and prosperous era. It is probable that Thales of Miletus was a trader of goods and ideas. He was a deep thinker and he had time to think, to observe nature, to learn of the wisdom of other cultures and to spread his views. Thales advocated that there were understandable regularities underlying the happenings of nature. He surmised that everything was made from one fundamental substance. If later records are correct, he correctly predicted the solar eclipse of May 28th 585 BC, which rather disturbingly occurred whilst the armies of the Medes and Lydians fought in battle. He diverted the flow of the river Hylas to enable a crossing by King Croesus. He studied natural clues to yearly patterns in the weather. One year he put his prediction of a bumper olive crop to the test by investing substantially in olive presses. As a result he became quite prosperous. Furthermore it seems that admiration for his accomplishments opened the door for Greek culture to actively encourage and support wisdom and learning. Thus this seminal Greek cultural era had begun.

So let me cherry-pick from the rich bounty of ideas proposed by subsequent generations of Greek philosophers. I shall start with

Anaximander, a pupil of Thales. Anaximander (611-547 BC) held that the fundamental universe was an indivisible whole that he called the apeiron. This apeiron had no restricting influences. Thus the apeiron was boundless with no notion of edges or breaks of any kind. In contrast, Heraclitus (535-475 BC) believed the universe was the creation of a permanent conflict. This conflict represented an eternal battle between irreconcilable opposites. Thus the universe was in a state of perpetual flux. Incidentally Heraclitus was a contemporary, on a different continent, of two other famous thinkers, Confucius (551-479 BC) and the Buddha (~563-483 BC).

Now, Parmenides (510-440 BC) believed there could not be such a thing as change or even spatial structure in any fundamental depiction of the universe. Democritus (460-370 BC) and Leucippus promoted the vision of a reality based upon copies of a single unique fundamental discrete unit form (the atom) moving in a void. Plato (427-347 BC) believed in a rational hierarchy built upon unchanging archetypical forms. These idealized forms could be justified by a self-authenticated first principle. Plato also believed that eternal truths (such as mathematical theorems) were more fundamental than knowledge of the flawed everyday world of appearance. In contrast, for Aristotle (384-322 BC), fundamental substance was an inseparable duopoly of quantifiable matter and qualitative form. Furthermore Aristotle placed an emphasis on the role played by an ultimate intended purpose for each thing. This supposed intended purpose was a fateful reason (the 'final cause') for nature of things.

After these great thinkers, innovation in fundamental ideas slowed. Some of the newer ideas were of a darker nature and societal support for philosophy waned. There had been earlier indications that Greek society was becoming complacent about the need to support learning. In 399 BC, Socrates (470–399 BC) was condemned to death. His supposed crimes were his non-orthodox Orphic beliefs and the corruption of the youth of Athens. However, the trial's motivation had more to do with the ambition, greed and power of established cliques. Perhaps Socrates educated the wrong people and thereby annoyed powerful people. The more learned might tend to know better what administrative goals to set but this skill does not make them better at achieving these goals in a cut-throat world of *realpolitik*. Anyway the era of independent Greek city-states was soon to end. In 338 BC, King Phillip II of Macedonia defeated the combined armies of Athens and Theba. This victory by King Phillip and his more famous son,

Alexander the Great, united and subjugated the Hellenistic world. This *status quo* remained until the Roman conquest of 197 BC.

Eventually the pious Roman Emperor Justinian (482-565 AD) snuffed out any hope for a recovery of the self-perpetuating intellectual vigour. In 529 AD, Justinian banned all teaching of Greek philosophy, which he viewed as paganism. Historians often use this singular event to mark the beginning of the European Dark Ages. Justinian's dictate illustrates the deep insecurity that all political institutions feel towards independent thought and any uncontrollable respected authority. Those who seek power seek to regulate and maintain order. They see freethinking as a subversive disruptive nuisance. We can only wonder where in the universe humanity might be now without this persisting irrational institutional fear of both learning and the pursuit of knowledge.

Now, order and stability marked these European Dark Ages. Thus this age allowed some cultural and economic growth. Indeed, had it not been for external disturbances like the calamitous bubonic-plague epidemics which (during the last millennium) repeatedly decimated Europe and Asia, we might still be living in the European Dark Ages today.

Some short-lived catastrophes can hold the seeds of a long-term blessing. Perhaps some people will applaud the fabled asteroid that, sixty-five and a half million years ago, may have wiped out the large dinosaurs and nearly everything else on Earth. This cataclysm opened an opportunity for smaller terrestrial mammals to dominate our planet. However, will the same people cheer again when the next asteroid hits? In a less dramatic manner the plague helped to weaken the influence of all societal dogmata. The staggering tolls of indiscriminate death brought into stark focus the need for a practical understanding of the way our world genuinely works. There was a difficult struggle for many centuries but eventually the European Renaissance flourished and the intellectual search for answers was reborn in a new discipline that later came to be called science.

Critically though, the rise of science saw the disciplined rejection of all 'mystical' concepts (ie those notions without clear empirical justification). Hence the clinical picture of the natural world (as traditionally portrayed by science) is limited to that which we can study quantitatively ie forms of matter and energy in space and time. Nevertheless, as we are about to discover, science is pregnant with further possibilities that are able to complete the dream of creating the final framework for a full understanding of reality.

Clearly all viewpoints are powerful constructs of ideas. However, often viewpoints are useful in some contexts and a hindrance in other contexts. Here I will draw attention to just two traditionally opposing mindsets from consecutive Greek philosophical periods.

The earliest period of Greek philosophy (ranging approximately from Thales to Pythagoras) was marked by a dominant science-like belief in building arguments supported by empirical evidence. The Pythagorean school of philosophy prospered at the end of this era and the start of the next. The role of number in the natural world – eg musical harmonies, architectural geometry and natural beauty – enthralled Pythagoras and members of his secretive sect. These natural associations led the Pythagoreans to argue that the rules of both mathematics and logic were incorruptible eternal truths. Hence the Pythagoreans held that these rules were both fundamental and necessary to the universe. Today most scientists probably hold this view. Many of these scientists rarely stop and ponder that a self-evident role for mathematics has no greater validity than any other empirically derived scientific law of nature.

Now, the period following Pythagoras saw an increase in the reliance on purely rational argument. These arguments became increasingly divorced from the presumed illusory world of experience. Parmenides was the most influential philosopher of this period. Many of his arguments start with the logical senselessness of non-existence. For instance, correctly stating that 'something' (that is elementary) does not exist is contradictory since there would be no 'something' for the statement to imply does not exist. Hence he argued that if a statement of non-existence is nonsensical, then logically non-existence is impossible. Therefore some elementary things must always have existed and the absence of all things is forbidden. A similar argument applies to any fundamental property that we define only by what the property is not. Thus we cannot have a void (ie nothingness). Consequently the Parmenidians concluded that the universe had no empty space ie it must be completely full. The Parmenidian school of philosophers went on to conclude that fundamentally there was no change, movement or variance in our universe because space was full. In their view, space was an unending unchangeable plenum. For the Greek philosophers these arguments proved overly influential; the arguments seemed to confirm a disturbing belief. This belief stated that everything in the imperfect world of our experience was an illusion. Arguments for and arguments against any proposition seemed possible. This unsatisfactory view led many philosophers to adopt a defeatist sentiment. Indeed a destructive suspicion, that ultimately we can prove nothing fundamental, still lurks in philosophical circles today.

Such despair is a product of rigid mindsets. It is easy to appreciate how mindsets like the empiricism of Thales, the rationalism of Parmenides and the later scepticism take hold and stick. These views are the core values professed by localized communities. Once an in-group accepts such fixed ideas, the sentiments are quite difficult to change or to reconcile. However, if the universe is at all rational, we have to be able to clear up the contradictions. This conflict resolution is a major achievement of eventual reality.

Imaginative Conspiracies

Let us leave the ancient world and turn to some more-recent bizarre ideas. We have stated that in this book we are planning to explain the nature of reality. So let us start with one possible definition of the term, reality. Reality is that which is both rationally believed and generally accepted. Now, many good conspiracy theories are eminently believable and often extremely entertaining. However, most of us do not entertain many of these far-fetched conspiracies because they presuppose a great deal, they make many unsupported conjectures and (since there are alternatives) they are an unlikely interpretation of the evidence provided. For a belief to be widely held we need few presumptions, reasonably unambiguous supporting evidence, internal self-consistency and compatibility with other rational beliefs.

I regret to inform some of you that Queen Elizabeth and the Pope are not members of a cartel that runs the planet's underworld. That easily damaged technology, underpinning non-stick frying pans, was not invented by incompetent inter-galactic aliens who crash-landed in the United States. The legendary singer and musician, Elvis Presley, is dead. NASA astronauts did land on the Moon on 20th July 1969, and they did bring back samples of Moon rock. Conspiracies give some people something interesting to believe. Presumably many individuals resort to such tangible conspiracies when, from their perspective, actual reality becomes intangibly weirder than a fictional story. Others may be cynics with a skewed emotional belief as to how the human world works. However, conspiracies also help to forge shared bonds in harmless eccentric social groups – the true believers.

Now, you may call those stories conspiracies, but perhaps we should not be too critical – they are quite unimpressive. Therefore let me show you what some fantastically magnificent conspiracies look like! Beware of the conspiracies generated by scientists and philosophers – the smarter the mind, the more ingenious the tale.

Thus a somewhat grander conspiracy involves the suggestion that our universe is a fraudulent imitation. This conspiracy theory conjectures that the knowable universe is a giant simulation – a way of experimentally investigating possible universes by a programmer in some other universe. Thus we are merely algorithms in a grand computer simulation. Hence within the simulation we are supposed to believe that we live in a real universe. This conjecture is entertaining and the ideas are adaptable to make several good movie plots. However, this conspiratorial conjecture describes an unlikely scenario for a very simple reason: all manufactured computers will have finite resources. Hence programmers must resort to computational approximations. Now, science examines the world so closely that the fraudulent nature of the universe should inevitably show up sooner or later. Furthermore algorithms may be smart but algorithms (as we know them) can only simulate intelligence. They are not alive. Unless you are willing to accept that you are only a zombie-like intangible piece of computer code, you can ignore this conspiracy. Of course if the programmer were to turn the computer off to go home at night, I would not be in a position to change my mind and neither would you!

Anyway, there are other interesting noble connivances. Thus the birth of modern classical (or Newtonian) science created several distinguished conspiracy theories. Apparently the idea of a Cartesian split between mind and matter seduced many philosophers. In this era it was accepted that science deterministically explained the motion of all matter. The Cartesian view was that everything (possibly excluding the divine) belongs to one of just two mutually exclusive classes – the mental and the material. This over-restrictive viewpoint led many philosophers astray into what we can perceive as an elaborate fantasyland. Some historians suggest that the problem started with a disciple of Rene Descartes (1596-1650), by the name of Arnold Geulincx (1624-1669). Some philosophers blame the Catholic Church for strongly promoting and possibly inspiring the Cartesian divide through the earlier teachings of Saint Augustine. However, blame just diverts attention away from the message to a messenger.

Descartes' clear mind-matter distinction posed a disturbing problem. Science said that matter had to move in a predictable manner. Thus there cannot be any possible causal interaction between mind and matter – matter is not free to respond to the immaterial mind. Consequently the Cartesian school of thinkers claimed that we must only witness a synchronous correlation between our decisions and our actions. Obviously this remarkable agreement between our thoughts and our bodily movements cannot be a random-chance coincidence. Therefore, the Cartesian school suggested that all the matter in the universe must have been set in motion from the beginning of time, according to the way the universe would inevitably unfold. Thus all matter subsequently moves (according to the natural laws of science) so as to give the illusion that our bodies are concurring with our personal decisions. Our decisions are quite free (so we can commit sins); however, a superior intelligence who manages the workings of the universe already knew (from the beginning of time) each decision that we are yet to make. Clearly this intrigue makes a good story and a truly wonderful conspiracy theory!

Intriguingly the eminent philosopher, Leibnitz (1646-1716), came up with a startlingly new conspiracy. He viewed the fundamental world as an infinite array of immaterial unextended monads. The word, monad, is used to signify that this is the *single* fundamental type of entity from which everything is made. The adjective, unextended, conveys the idea that these monads have no spatial size; monads apparently reside in space but they do not extend over any region of space. This monadic picture is interesting because it does slightly resemble the true nature of reality that we shall be describing shortly. However, the nature of the Cartesian divide between mind and matter grossly misled Leibnitz. Hence Leibnitz supposed that if monads were not matter, then surely they must be of the alternative mental and spiritual realm. Thus Leibnitz decided that the immaterial tag (that he had given to monads) meant that monads were the immortal elements of souls. In addition, he took Geulincx's ideas to their full conclusion. If minds and bodies are both made from monads, and minds and bodies cannot interact, then monads must not interact either. Thus in Leibnitz's magnificent conspiracy of universal proportions, monads were predestined to behave in a programmed way to give an illusion of the synchronization of all actions, all interactions, all experiences and all thoughts. Wow! One lesson is clear from this conspiracy. Even exceptionally bright minds can go superbly astray when exploring what are almost sound basic ideas. Anyone may start with good ideas but figuring out what the ideas imply is quite another matter.

Can we top that? Well another distinguished philosopher, Bishop George Berkeley (1685-1753), after whom the university town of Berkeley in California is named, took another conspiratorial perspective. Berkeley rejected the dualism of mind and matter. He argued that the mind existed but he had no sound evidence for matter. It made little sense for there to be two fundamental forms. Hence it was safest to believe that the material world did not exist and the material world was merely a convenient illusion formed from mental ideas and made to appear real by the consciousness of the divine conspirator general, himself. This is surely a beautifully contrived conspiracy suited to the beliefs of Berkeley.

However, it seems that not all old conspiracies die; some conspiracies just get more preposterous. Hence today you still find philosophers who think that they like us are misguided dupes. Apparently some modern philosophers maintain that we do have conscious experiences, we have feelings and we understand many things around us. Nevertheless they do not believe that we have any free will. The idea of free will is apparently a convenient delusion that helps you explain to yourself why you are reading this book. Nature has conspired to imprison us within material bodies that are committed to fulfilling the actions of a lengthy and sophisticated play entitled the conspiracy of entombment. Next time you speak, you are just reciting the next line of the script. Even if you hold your tongue, that action is what has been preordained. If this thought demoralizes you, then that must have been the intention of the playwright. You can have no responsibility for your actions. Neither do I bear any responsibility for having told you. I think that I have written the lines of this book, but they are in the original manuscript of the play. If you can, you should relax and enjoy the play. Of course if you can force yourself to accept this conspiracy you might join the extreme ranks of those who also believe that unconscious zombies populate the world. Zombies behave and talk exactly as if they were alive. Zombies have no feelings, so their suffering is irrelevant. I think proponents of these views must be joking (or seriously misinformed) when they argue that *they* are taking science seriously.

Now, changing the whole concept of a conspiracy, it is fair to ask: Could there be a universe with no reality (ie nothing that is rationally believable)? Clearly there is a reality that we experience; so what we are asking is whether things could be otherwise. If a (rational) reality did not exist, we would have no need for logic or statements. However, our earlier definition of reality surreptitiously introduces the concept of logic. Thus suppose nothing else (other than this statement) was rationally believable. Then, we would at least have the previous statement as a virtual rational belief and hence we would have a reality of sorts. This denial of negation is an old argument and one that appears to be a play on words. It reflects a problem we have with Pythagorean thinking that views the world as number. When we view the world as quantitative, then nothing and something are always mutually exclusive. However, when we view reality as qualitative, then nothing is a particular something, an absence is relevant knowledge. Hence it seems there always needs to be something that is rational, that exists, etc. Maybe that argument sounds conspiratorial too.

It is also fair to ask: Could there be many alternative realities? In principle perhaps there could be many alternative sub-realities. However, the totality of these sub-realities comprises the reality of our provisional definition. Thus we find that our definition of reality ensures that there is one reality. Shortly we will generalize this definition and remove hidden presumptions.

I mention this alternative-realities consideration because it is relevant to a popular modern conspiracy theory called the multiverse. Like many conspiracies this conjectured belief is fuelled by an admirable desire to understand the world. However, ultimately the conspiracy fails even to achieve this aim.

There are many versions of this conspiracy, but let me pick just one extreme example of them. Hence in a version of this conspiracy, the multiverse (ie universe) comprises an indefinite number of independent alternative monoverses (ie sub-universes). Each monoverse is a distinct universe-like entity in its own right. The conspiracy says that we can place no constraints upon the form that any monoverse can take. The nature of reality, the laws of physics, the dimensionality of the universe, the constants of nature, etc are set at random within each monoverse. The monoverse that we inhabit is special, but the reason is that this monoverse supports life like us.

Let me explain this remark. Recall that we live on a rocky planet with liquid water and an oxygenated atmosphere. This location is not typical of the universe, which is mostly almost nearly empty space at a temperature of -270.4 °C (2.7 Kelvin). However, the fact that we did not evolve in the cold reaches of outer space is unsurprising. Our bodies could only have evolved in Earth-like conditions. Thus it is no surprise that our planet has Earth-like conditions. We call the general principle behind such reasoning, the anthropic principle.

Well, we use a similar anthropic argument for explaining the observed features of our monoverse. Most monoverses would not support or evolve those forms that we know as life. Thus out of all the myriad of monoverses, our monoverse needs laws of science like those that we observe so that we could be found here.

This multiverse hypothesis is an uninformative conspiracy because it says that there is no fundamental reason for anything. Nothing has to be exactly as it is. Everything could be different. Indeed everything actually is different in some other monoverse. Most of these alternative monoverses are examples of pure nonsense: they have no laws of nature, no mathematics and no logic. Thus an infinite collection of undetectable monoverses conspires to provide an accidentally observed universe. We can never know that these other monoverses exist, yet their existence is supposed to explain why we see order and sense in our observed monoverse.

Whilst this conspiracy may offer an unsatisfactory explanation for our monoverse, the argument fails to explain anything about the multiverse. Where does the multiverse come from? In fact the conspiracy tells us that studying our monoverse offers no insights into the infinitely variable universe. Thus the multiverse hypothesis generates a giant stop sign for science. The sign says that we have reached an ultimate limit of understanding. We can learn no more, so we should stop trying. This unwarranted speculation would be a staggering let down, if it were true. Our universe is amazingly understandable. It would be quite perplexingly frustrating if we could never understand why this is so.

Furthermore the infinite variability of monoverses conduces to some quite profound problems. Lawless monoverses would have to be the most common. Hence statistically we are likely to be in one of those rare irrational universes that (like the universe of fabled monkeys accidentally typing the complete works of Shakespeare) have been just pretending – up to now – to be rational.

This leads me to my final provocative tale. Thus let me end this section with the crazy conspiracy of late-determination. The idea from which this belief arises is that, at the start of time, there is nothing to fix the laws of nature. This may seem reasonable, but somehow we are being asked to believe that time exists before there are any laws to require anything like time!

Anyway, the story asserts that initially the laws of nature are entirely fluid (ie undecided). Now, this is not a multiverse theory. There is only one universe, which is tentatively in an ill-determined (entangled) quantum-like state with respect to uncertain laws (like quantum mechanics). Remember that conspiracies do not have to make good sense. They just have to be told well – and I am not going to do that. Thus supposedly the universe has no decided laws until one fateful day. On this day the universe conspires with life in general. The universe potentially evolves a certain magnificent life form. This magnificent life form joins the conspiracy and makes a series of measurements to evaluate the universes' laws. Thus the quantum-like state of the universe instantaneously collapses into a state with laws of nature seemingly fixed from the very start of time. These laws of nature do not just allow the magnificent life form to have a real existence in the universe. The laws effectively require that this magnificent life form must exist so that it can make its measurements and get particular results from these measurements. Thus this most significant sentient creature in the universe has conspired (after the fact) with the universe to create a universe fit for itself. Then, the rest of us have to put up with living in a few leftover isolated rocky nooks and crannies within someone else's universe – a universe created by and for a space-alien. Our universe-selecting life form could still have got slightly different answers for the physical constants. This is all most unsatisfactory. Unfortunately we cannot know if other life forms might exist with significantly different laws of nature. We cannot even know who precipitated and hence perpetuated the conspiracy in the first place.

Rise of Science

Conspiracies aside, what further ideas did the modern era add to the mix? Well let me date the start of the modern era from the conclusion of Magellan's momentous naval expedition (of 1519 to 1522). This expedition completed the first circumnavigation of the globe by sea. Magellan himself did not survive the trip but his fleet's legacy could be likened today to establishing a permanent outpost upon Mars. (I do not see Mars as a second home for humanity; I see it as becoming a pest-free food bowl. Water and sunlight might be in short supply; however, after a little terraforming, we could cheaply grow vast acreages of pesticide-free crops.) Such historical events just are unforgettable and irreversible. They are like those few flakes of snow that start to roll down a slope; they slowly grow in size until they trigger an avalanche. Now, as with the short survey of ancient-Greek ideas, I shall be brief and highly selective in collating this review of the scientific ideas from the period that followed Magellan's last voyage.

We start our rapid review of the big-picture advances in the sciences with Francis Bacon (1561-1626). Bacon looked at the methods of science; he formalized and promoted the method of induction as a legitimate method for the development of science. Meanwhile his contemporary, Galileo Galilei (1564-1642), recognized that the legacy of Greek thinking needed substantial revision. Christianity had adopted Aristotle's central role for purpose. Thus the church taught that things and people had a rightful place. Naturally these things made an effort to move towards their rightful place. Heavier objects likes stones were more driven and hence they obviously fell more quickly than lighter objects such as feathers. Galileo recognized the fallacy of this picture. He identified the role of air resistance and even buoyancy. He argued in a work of fiction that, in vacuo, all bodies would fall at the same rate. He also acknowledged the sensibleness of the Copernican view that the Sun was close to the centre of the solar system. Unfortunately Galileo was in the wrong place in Europe to hold such revolutionary views.

Let me briefly digress from this scientific survey. At this time, Italy might have looked to be a likely place for a rebirth of a pursuit of rational free inquiry. Certainly honour was awarded to the creative works of Leonardo da Vinci (1452-1519) and Michelangelo Buonarroti (1475-1564). The peaceful ambience, prosperity and confidence of Italy with its city-states and the tradition of patronage almost made it an obvious society for awarding respect for reason and the free debate of new ideas. Furthermore the fall of Constantinople in 1453 had led to an influx of Byzantine Greek scholars with new viewpoints on the world. It seems that there was just one hurdle. This obstinate obstacle guaranteed that the major prizes of leading the discovery of new scientific knowledge, of leading the industrial revolution and of founding the predominant world empire would bypass the entire Catholic world. Even today, the dark shadow of this terrorful injustice still haunts the recesses of European minds. Thus with this spectre in mind, the *Monty Python* team made light of the dreaded barrier to progress with the cheery refrain: *Nobody expects the Spanish Inquisition*. Hence the renaissance of reason was forced to occur beyond the shores of Catholicism.

Let us continue our trail of mankind's accomplishments. Elsewhere in Europe, Rene Descartes (1596-1650) felt freer by leaving Catholic France. In his work, he exposed the connection between algebra and geometry but his most lasting and divisive contribution was to argue that mind and body were very distinctly different things. I have already made several remarks about this division.

Back home in France, Pierre de Fermat (1601-1665) was content to study less controversial issues. Today we often remember Fermat for his mathematical theorems and his work on the theory of numbers. However, he also showed how the trajectory taken by a ray of light always corresponded to the quickest path between the starting and ending points. This is a small idea that appears insignificant. However, it is like the tiny acorn that grows into a mighty oak. Hence I shall be discussing the implications of this observation in detail later.

This intellectual climate prepared the way for a coming major advance in understanding the universe. Respect for knowledge and reason were returning to centre stage in Europe. One could not have predicted precisely when or where the critical breakthrough would come. Yet it was indubitably coming. Actually the great plague of 1665-6 set the timing. This plague epidemic was the last widespread outbreak of bubonic plague to decimate Europe. As a strategy to minimize plague deaths, authorities closed the British universities. Hence a young yeoman, named Isaac Newton (1642-1727), returned to his rural home, a sheep-farm in Woolsthorpe, Lincolnshire. Newton's father had been a farmer. However, his father had died before Newton was born. Newton had acquired no interest in farming. Thus for the next two years, Newton had abundant time and minimal distractions. It was during this time, possibly reclining under a shady apple tree, that Newton accomplished some of his finest work. His unrivalled success was to transform the ethos of British society. A spirit of achievement – born on a tidal wave of scientists, engineers and inventors – transformed the world. Such a tsunami is unstoppable. Thus Newton fashioned many of the subsequent events of world history. Undoubtedly his accomplishments even contribute to the reason why the language of this book is English.

What was so remarkable about Newton's revelations? What did Newton achieve? Well, Newton transformed the structure of natural knowledge into a tool capable of predicting the quantitative behaviour of matter throughout the cosmos. He created the first consistent scientific picture of reality. Nobody born since Newton's time can really appreciate just how momentous his contributions were. Because of our education, it all seems so obvious to us now. In this worldview, we picture the universe as a system of interacting corpuscles (ie small material bodies). These corpuscles were set in initial motion and ever since have complied with the commandments of precise laws of nature. These laws describe how matter moves under the influence of various forces acting between the material bodies separated at a distance. Today we know these laws as Newton's laws of motion. Newton confirmed that the world was mathematical, as predicted by Pythagoras. He showed that the same laws of nature (eg Newton's law of universal gravity) applied to the orbit of the Moon and the fall of an apple on Earth. Practically overnight, Newton turned the mystical proto-science of the ancient Greeks into a fully-fledged workable science. Such a transition is unrepeatable. However, even Newton did not perfect classical mechanics. He left a little work for succeeding generations. Hence let us continue with the survey.

Joseph Lagrange (1736-1813) was able to improve the formal structure of Newton's classical mechanics. He demonstrated that one could derive the laws of classical mechanics from the stationariness of a quantity that we call the Lagrangian. Minimum and maximum conditions are examples of what we call stationariness. Lagrange also developed the method of Lagrange multipliers for formally solving constrained variations. However, not everything was entirely rosy. Pierre-Simon Laplace (1749-1827) disturbingly argued that the predictive power of science seemed to rule out free will for all matter. Furthermore he also noted that the inverse square law of gravity could be interpreted as a solution of a simple second-order differential equation that we call Laplace's equation.

In this way, over time, new thinkers pushed forward at the frontiers of comprehension. William Hamilton (1805-1865) looked for a way to generalize complex numbers, which denote two-dimensional points on what we call an Argand plane. Hamilton looked for a number system that could perhaps describe three-dimensional space. Eventually he discovered a new system of numbers known as quaternions. These numbers give an early indication of relativistic space; quaternions actually describe a four-dimensional space. Then Charles Darwin (1809-1882) and Alfred Wallace (1823-1913) created the paradigm of evolution. They formulated their theory of natural selection that provides for the preferential survival of the so-called fittest. Then James Clerk Maxwell (1831-1879) unified the fields of electricity and magnetism. Furthermore Maxwell developed the several connections between microscopic and macroscopic interpretations of matter. For instance his work on perfect gases led him to a formulation of classical statistical mechanics.

It seems that science was making its final touches to the great achievement of classical mechanics when along came the twentieth century. Max Planck (1858-1947) discovered that the only way of reconciling the spectral form of blackbody radiation with other known science was to require quantized emissions of radiation. Then Albert Einstein (1879-1955) stunned the world with his conclusion obtained by using rational arguments about the lack of special places, special velocities or unique interpretations. Thus Einstein reformulated classical mechanics in relativistic terms. He replaced Newton's awkward absolute space with a consistent relative space-time. Hence he combined mass with energy and spatial geometry with gravitation. The theory allows distant observers different appreciations of the same happenings. We should view this as improving classical mechanics. However, Einstein also showed how quantum principles applied to the photoelectric effect. Something else was brewing.

The bloated paradigm of classical mechanics was strained and about to explode. Planck had shown that we needed to make some changes. Niels Bohr (1885-1962) pointed out how Planck's quantum ideas could explain the structure of the atom. Then Werner Heisenberg (1901-1976) gave to us the first approximate expression for the equations of quantum mechanics. This result was soon followed by a wave formulation developed by Erwin Schrödinger (1887-1961). Paul Dirac (1902-1984) helped initiate many further developments of the formalism. In particular Dirac is known for adding special relativity (the Dirac equation) which predicted a particle property called intrinsic spin and predicted anti-particles. However, many classical features (materialism, Newtonian definitions of quantities) passed unaltered into the theory of quantum mechanics. Meanwhile Carl Jung (1875-1961) drew inspiration from these so-called counter-intuitive developments in physics. He carefully reassessed, analysed and categorized the styles of thought used for the mental processing of information.

This is as far as I go in my survey. Clearly in any quick survey there are many omissions. Most likely, your favourite scientist is absent from my list. For instance, I have not mentioned maestros like Mendel, Pasteur, Mendeleyev, Lord Rayleigh (John Strutt), Fleming, Rutherford and Faraday. Certainly such scientists excelled in uncovering important facts and presenting their interpretations. However, in this work, we are more interested in exploring far-reaching ideas with implications for fundamental knowledge.

Now, I am reminded that some respectable authors sometimes add drama to their accounts by selective quotation. It can be entertaining to quote pre-eminent scientists. We can choose between many quotes that imply that nobody really understands a theory like quantum mechanics or general relativity. However, such quotations can be glibly misinterpreted. The actual sentiment is more subtle. The theory did not turn out as the scientist expected. The scientist fully accepts that nature is behaving in this way, they understand what the theory implies but they cannot see why the behaviour is contrary to their expectations, they suspect that something may be missing. This surprising consequence nearly always happens when we have to unlearn something that we have previously taken for granted. The new theory seems counterintuitive because of a long familiarity with previously successful superseded ideas. These magnificent quandaries make science so fascinating because they allow us to lift a veil and stare a little deeper into nature's endearing mysteries.

As I mentioned, I am stopping my summary here, although I do feel compelled to mention a rival theory in the next section. To ignore almost entirely the achievements of the last 50 years may seem harsh. Certainly we have learnt many significant scientific facts. We know a great deal more about elementary particles, about the macroscopic cosmos and about the chemistry of life. We have a range of hypotheses to combine quantum mechanics and general relativity and even imaginative suggestions as to how to heal the mind-body rift. We have a mapping of the genome of several life forms; we have achieved mammalian cloning and developed all sorts of associated biotechnologies. Yet perhaps the importance of all this work is only to show that we need the hypothesis of eventual reality or something like it. These achievements are barren without a comprehensive underlying scientific rationale. I would like to suggest that only a theory on the edge of science can provide this rationale.

Surely it's M-Theory?

Have you heard about M-theory? Some folk would have you believe that M-theory is the last great hope for a final solution to everything. They believe that M-theory is the fabled theory of everything. Of course it is not even the only mainstream contender. Numerous physicists are working upon many competing formulations. However, M-theory gets much of the publicity. Now, I have to admit a bias against M-theory. M-theory has little to do with the hypothesis of eventual reality and I have only a cursory knowledge of the theory. Nevertheless let me give a simplistic and unashamedly distorted assessment.

We shall start with some history. There were stunning advances in theoretical physics during the early twentieth century. We saw the advent of special relativity in 1905, general relativity in 1915 and quantum mechanics in 1926. Then we saw physics undergo an improvement and consolidation phase. Our pictures and understanding of the universe were refined but not replaced. We can all understand what destructive World events interrupted this progress. Anyway towards the end of the twentieth century some physicists began to get restless. Where was the next major advance? Some commentators began to talk hopefully about an impending theory of everything. Since physicists viewed the world as describable by number, these commentators meant a theory of all those things that were quantitative. This is why I do not proclaim eventual reality as a theory of everything. Everything quantitative is but a small part of the scope of eventual reality.

Let us consider more background issues to the structure of matter. The idea that matter was ultimately made of interacting point particles is older than Newton. Indeed we normally attribute this idea to the Greeks, Leucippus (485-420 BC) and Democritus (460-371 BC). They argued that if the building blocks of matter had size, then we could further split this matter into smaller parts. Thus we cannot regard bodies with a finite size as being the fundamental constituents of things. Everything had to be made of point particles called atoms.

In contrast much of twentieth-century physics suggests that nature does not accept point particles. For instance, whenever we combine quantum mechanics with special relativity, unwelcome infinities crop up in the analysis unless we impose a minimum spatial resolution. Even from classical considerations we can see how these problems arise. Imagine an electron as a uniform shell of negative charge. As we make this shell smaller and smaller, the repulsive force of the 'bits' of like charges increases. This is a consequence of the familiar (inverse-square) Coulomb force law. Thus we have to do more and more work to squeeze the bits of charges closer together. Hence if we could shrink the shell to a point, then the self-energy would become infinite. Furthermore general relativity also showed that the gravitational aspect of macroscopic matter had a finite size. Perhaps you have not heard of the Schwarzschild radius but you have probably heard of a black hole's event horizon – matter may only cross this surface in one direction. Thus if you venture too near to a black hole you cannot return because the character of space changes on the other side of the event horizon. This effect gives the minimum-size scale of macroscopic gravity. Then, combining Heisenberg's uncertainty principle of quantum mechanics with general relativity, we discover that even the idea of space, on an ultra-small scale, becomes quite uncertain. We call the very small scale on which space becomes meaningless, the Planck scale. Thus there were plenty of clues that the hierarchy of matter does not shrink down to a sizeless point. There had to be another answer. Therefore people went looking for other possibilities.

Now, the known hierarchy of matter starts in space with the universe and descends through various galactic structures, then down through galaxies and stars. This hierarchy continues with planets, rocks and molecules. Then, molecules are made of atoms. Atoms are made from electrons and nuclei. Nuclei are made from protons and neutrons. We have to ask the questions: Does this hierarchy go on for ever with particles made from more fundamental particles? Alternatively, at some level, do we find a final layer of particles that are made from something else or even nothing at all?

Well, neutrons and protons are made from quarks. However, quarks are rather weird. They look like particles in many respects but they are subtly different. Here are the first clues possibly showing that we have reached the end of the line for building matter from smaller units of matter. We can see quarks in fuzzy images within say a neutron. However, we cannot take a quark out of this neutron to weigh it on a pair of scales. No one has ever seen an isolated quark. Indeed we believe that no one can ever see an isolated quark. Quarks are necessarily trapped within clumps of quark matter. Secondly quarks have an extra quantum number called hadronic colour. Hadronic colour is not possessed by any observed particle. Hence one can only put special "colourless" combinations of quarks together to make observable particles like neutrons. Thirdly when we look at the experimental masses of some excitations of these observable particles, we see evidence of a simple interaction between the constituent quarks. Hence the idea was born that the quark-quark interaction was like a string of energy.

The question was asked: Why put quarks at the end of these strings, why not make quarks *the* ends of strings? Of course, as string theory developed, the theory became reinterpreted and reinvented several times. These later much publicized 'revolutions' have been described in several popular books. Now, the theory has multiple extra dimensions for space, super-symmetry, closed-loop strings and string vibrations. Yet despite all the changes, the theory is still not quite right. Therefore to increase the options, this picture of one-dimensional strings was generalized to include higher-dimensional things called branes.

As this theory continued to develop, there were numerous hold-ups, breakthroughs, twists and turns. Undoubtedly there is some truth in some aspects of the model; however, this does not make the theory valid. Perhaps the theory's power comes solely from its complexity.

Given the enormous difficulties encountered, it has an impressive record of endurance. Yet this is also the theory's weakest feature. In the past, whenever we have made an important advance in science we have added a tremendous boost to our understandings. At first, the new theories might have seemed difficult to accept but they were inherently simpler and more explanatory. The increase in simplicity, scope and power has meant that eventually we can find illustrative experiments to test the theory. However, the still imprecisely specified M-theory is just adjusting itself to fit reality. Any justification for the way things are is far beyond its reach.

Hence the analogy that I draw to describe this progress is with a high-power sports car named physics. The driver negotiated some tricky bends in the early twentieth century and then stopped for a much-needed rest. The Germanic cultural climate that had encouraged so much of the fundamental thought (of the early twentieth century) was broken. However, as a newer slower driver resumed the journey he mistakenly took a dirt track (sign-posted to M-theory) running alongside the main highway. After a while the dirt track petered out and the ground became soft and sandy. Nevertheless the driver was ingenious; he let air out of the tyres so that the tyres could get a better grip. Every now and then, the car became bogged and the car had to be dug out. Even now the car continues to make slow progress in an uncertain direction. To my mind, this slow progress is a clear indication that this theory is on the wrong track. Perhaps you do not agree, but physics is intrinsically simple. It needs to be; only people complicate it.

Now, this analogy is not a denigration of the highly proficient work undertaken by many of my colleagues. Rather it presents a cautionary suspicion. It is reasonably common for science and metaphysics to go too far along a particular route. Recall Leibnitz's theory of monads. Inevitably we accept some ideas because we suppose there are no other acceptable alternatives. We are egged on by earlier successes, and we forget the provisional tag. However, after awhile, credulity is stretched too much. Like an overstretched rubber band, the inevitable consequences are dramatic.

Natural evolution sometimes follows a similar route. A species can become overspecialized and hence becomes an evolutionary dead-end. Thus the extant genus can effectively retreat and take another course. Is this the situation today with M-theory? Well, we observe many symptoms of an ailment: umpteen unexplained dimensions wrapped up on an ultra-small scale, generations of never-seen particles, impossible looking mathematics, a surfeit of formal possibilities and no near-term likelihood of a fundamental explanation. Thus eventual reality happily abandons these recent accomplishments and connects with the physics of simpler times. The resulting hypothesis is simplicity reincarnate.

2. What Do You Mean by Mean?

Well, what do I mean by mean? This is a most dangerous and subtle question; one may even say that it is loopy! However, because of its confused nature, I will not answer this question directly. Yet, I hope that you will be able to see how to answer the question for yourself by the end of the book. With this objective as an aim, I will begin the task for you in this chapter. We will decide what is meant by some other terms.

The courageous attempts by some of history's most notable philosophers and scientists – to clarify aspects of fundamental reality– led me inconsiderately to ridicule some of their conclusions in the last chapter. Let me say that I did not mean any disrespect for the pilloried philosophers and scientists. Only a brave man or woman boldly ventures forth into the territory of new ideas. Thus I sought to entertain by recalling some earlier ideas that were possibly formed at a time when we knew much less about the world. It is to our credit that we are all human and fallible. Furthermore, in this work, we are about to venture into this treacherous territory also. I hope that you find this exploration even more exhilarating.

Clearly whenever we transgress from the rigorous realm of science into the dispute-ridden areas of philosophy, we meet disagreements over the meanings of words. Inevitably some semantic arguments are quite senseless. Thus even distinguished philosophers have been known to be so disenchanted with their brethren that they publicly accuse their colleagues of preferring to use words as tools to impress audiences, rather than elucidate ideas.

Therefore what is the real issue? We all probably think we know the meaning of the majority of common words. Nevertheless this belief can sometimes get us into trouble. Certainly we all possess a native language, a language in which we are comfortable reading, writing, speaking and thinking. However, dictionaries reveal that most words have multiple meanings. Even ignoring such ambiguities, could you satisfactorily answer the questions: What is wisdom? What is time? What is thought? What is reality? We commonly use such linguistic terminology; we can describe some attributes that will identify these terms (ie wisdom, time, thought and reality). Yet when we come to ask the critical 'What is' question, most people (those unfamiliar with eventual reality) are stumped for a definitive answer.

Readers, who have not thought greatly about this, might suppose that there is no problem – we know what we mean. However, if this were really the case, then they would know how to get answers to questions such as: At what stage during our physical development do we take our first thought? What is the reason for universal existence? Are there limits to wisdom? Could we make a thinking machine? Could we make a wise machine? Could we go back in time? Why do we experience time flow? What are the fundamental elements of reality? Can plants think? Could there be hyper-intelligent space aliens? These questions show that perhaps we do not know what these selected concepts truly are. Moreover, to uncover the fundamentals of reality, we will need to talk about many of these seemingly undefinable basic things.

Nevertheless this conventional omission is not too much of a problem. After all, when it comes to word usage, words just mean what the originator intended them to mean. Therefore it is incumbent upon me to try to express myself clearly and let you form your own understanding. Now, there are certain basic concepts from which we are going to build our understanding of everything. Indeed to appreciate the hypothesis of eventual reality you will need to understand what I mean by the terms: reality, natural philosophy, science, number, space-time, logic, mathematics, matter, mind and thought. Hence within this book, most of these concepts will be clearly explained and then defined. Thus let us make a start with our first explanation and definition.

Reality

To introduce this discussion on the topic of reality, let me take you back in time almost two and a half thousand years to ancient Greece during the time of Plato. Perhaps the most famous of Greek philosophers, Plato, was born to a well-to-do ruling-class Athenian family around 427 BC. Had he been born outside the Greek world, say in Britain, it is likely that he would have been a Druid – a learned man entrusted with maintaining cultural customs and traditions, dispensing medicines and interpreting heavenly signs. However, Plato was fortunate to be born with a significant intellect into a society that prospered because it had generally prized freethinking, rational debate and education. Furthermore Plato had access to a wealth of earlier Greek philosophical thought. Historians generally suppose that Plato was a student of Socrates (470-399 BC), although Plato's privileged upbringing may have placed him as a casual client rather than a member of an inner circle of pupils. Certainly Plato was quite incensed by the manner of Socrates death. Later when aged about forty-seven, Plato set up and headed the Western hemisphere's first institution of higher learning, known as the Academy.

In creating his worldview, Plato brought together and expanded upon many conclusions of earlier Greek philosophers. Thus he accepted Pythagoras' views on the primary role of numbers in the world, the divine nature of eternal truths such as mathematical identities and laws of nature, and a belief in an unperceived ideal world somehow lying beneath the world of direct experience. He accepted Heraclitus' claim that the structure of the perceived world was in a state of permanent flux ie everything was in motion. Yet he also accepted Parmenides' arguments that ultimately spatial and temporal variations were illogical, thus there could be no change and no void in an underlying fundamental world. We cannot detect or sense this hidden fundamental world in any experimental way. Thus this underlying world is only to be understood by pure intellectual endeavour. Hence, with these shared beliefs, Plato developed the view that the fundamental basis for reality was constructed from universal archetypical forms – forms that epitomized beauty, perfection and constancy.

Now, it seems that all great thinkers are associated with at least one gedanken experiment. A gedanken ("thought") experiment is a hypothetical experiment that one just imagines carrying out. By applying a little logic and commonsense we can reveal useful informative without having to conduct (often impractical or even unethical) experiments. Thus gedanken experiments are very useful devices and I shall be mentioning several gedanken experiments. The first such experiment that I shall discuss is Plato's cave.

Plato imagined that a group of imprisoned people were reared (all their lives) in a cave. Chains shackle these unfortunate wretches. The chains even constricted the movement of the prisoner's heads. Thus all they can actually see of the world around them is a cave wall. On this wall they observe the shadows of their captors, of animals and of artefacts being used. This wall also reflects all the sounds that they hear. Plato suggests that such constrained prisoners would develop a misguided viewpoint. He suggests that they would reasonably form the opinion that the shadows were their captors and the events shown by the intangible moving shadows were real events. Plato uses this argument to illustrate his view that the physical world that we see around us could likewise be an illusion. Plato proposes that there is another unseen world from which our observable world is constructed. Clearly experiment probes the seen world. Thus this unseen world is accessible only by philosophical inquiry.

Now, in Plato's tale, the prisoners believe that the shadows are real constituents of their surroundings. It would be inappropriate for us to condemn the prisoners for forming these views. In the same circumstances, we would probably do no better than they would. Hence let us adapt our language to reflect alternative (but related) views. An accepted worldview (ie a system of beliefs) confers the status of a reality and the notion of what is real. Thus shadows are indeed deemed real within the worldview of shadow-land. Hence this allegory of the cave illustrates how we actually decide what is real. Reality is simply an attribution of a worldview. Consequently we need to make clear what worldview we are using when we refer to real and reality. The prisoners correctly believe that the shadows are shadow-real whereas their captors would recognize that (in their worldview) the shadows were simply a trick of the light.

It turns out that, in science, we have already faced a similar situation – a radical change in worldviews with the older worldview subsumed by a newer worldview. Sir Isaac Newton drew up the first comprehensive scientific worldview. His worldview was developed into the classical worldview that prevailed until the early twentieth century. In this picture, the world is made of material point particles that move in a given space. These particles experience forces at a distance. These particles have a definite existence. A set of quantitative values such as position, velocity, mass, electrical charge, etc completely specify each particle. Furthermore the quantitative properties of these particles obey clear scientific laws that predict what and how things change in time.

Every reader should be familiar with most aspects of this classical worldview (ie beliefs of classical mechanics). We have all been taught the classical worldview at secondary school as though the classical worldview were the commonsense view of reality. However, we should not accept that classical reality is especially commonsensible. Surely if the classical worldview coincided with commonsense, then the ancient Greeks would not have devised their alternative mystical beliefs. Now, I do not wish to dwell on arguments about what is commonsense. It is more important to note that the classical worldview is ingrained in most people's thinking. This is unfortunate. We do have difficulty giving up the fundamental classical notions once the educational system has taught them to us. For instance these classical notions include ideas of certainty that physicists have *not* accepted for over eighty years. Indeed I find it quite shocking that only those who do a university course in a physical science have any genuine exposure to the almost century-old breakthroughs that created the quantum worldview.

In this quantum picture, the universe is made from quanta that move in a given space. There are just two overall styles of quanta. Thus the known elementary particles (such as electrons) are a special form of quanta known as fermions. Indeed nature constructs everything found in space, from assemblies of unit quanta. For instance electromagnetic waves (eg light and radio waves) are made from quanta known as photons. This type of quantum is known as a boson. Classically we might identify fermions with point material particles and bosons with energetic waves of radiation.

However, this distinction is not as useful in the quantum realm. Fermions can behave like waves, and bosons can behave like particles. Waves and particles are characteristic features of the classical worldview not of the quantum worldview. One way of demonstrating this more general quantum view is to consider a cylinder. Look at the cylinder end-on and you see a circle. Look at the cylinder side-on and you see a rectangle. However, the cylinder is neither a circle nor a rectangle. The cylinder is not any two-dimensional shape; it is a three-dimensional shape. Likewise the quanta are not particles or waves. The term, wavicle (a contraction of wave-particle), was coined to describe the generalized quantum form. Nevertheless we do not use this term often, as the wave-particle duality is rarely an issue to quantum practitioners.

Now, there are no forces in this quantum picture – forces are only classically real. In place of forces, overlapping quanta interact locally via a coupling. Furthermore the definite certainty of classical reality is gone. If we precisely determine one element of classical reality, say the momentum of a quantum, we cannot have any knowledge of its complementary property (ie in this example the position of the quantum). Let me use the cylinder analogy again to illustrate the point. If we insist on viewing the two-dimensional circular face of the cylinder, we cannot simultaneously see the two-dimensional rectangular view. Position and momentum (for fermions and bosons) are quantum-complementary variables. The more accurately we specify either of the classical variables (position and momentum), the less precise becomes our knowledge of the other classical variable. This is the basis of the Heisenberg's uncertainty principle.

A set of quantitative values completely specifies a system of quanta; however, the appropriate set of labels depends upon the state of the system. Furthermore clear scientific laws (that predict what and how things change with time) define the temporal evolution of a system of quanta. Nevertheless, when we make a measurement, we force the system to display a classically real property value and then we can only predict statistically the outcome of the measurement of the property. This is our picture of quantum reality. Clearly quantum reality is very different from classical reality; nevertheless we can build features such as classical definiteness from the statistics of systems containing many quanta.

Now, even though quantum reality is more basic than classical reality, it is still useful to talk of a classical reality. As with the example of Plato's cave, the worldview assigns the description of reality. Hence we shall refer to a quantum reality and a classical reality. These realities use different concepts and at times seem to be at odds. This difference is how we generate so-called paradoxes. However, nothing is ever really mysterious – it only seems that way if you have only been taught old science.

Thus so far we have accomplished two things. Firstly we have tightened our conditions on the use of language associated with reality. Whenever there may be ambiguity, we shall require the provision of labels to identify the prevailing worldview. Secondly we have transferred all the questions about the nature of reality into questions about worldviews.

Hence we now ask: Are we to accept just any worldview? Are we to accept only scientific worldviews? What is science anyway?

The last question (about science) is interesting and we shall be considering it in detail shortly. The second question (about acceptable worldviews) is easy to answer. It is clear from the work of the Greek philosophers that we ought not to restrict our interest only to the scientific worldviews. Indeed, in this work, we will also be interested in any relevant and comprehensive worldviews provided by natural philosophy. Natural philosophy encompasses science, so we shall look to substantiated worldviews, taken from natural philosophy, for our hierarchical worldviews.

Furthermore I see no linguistic difficulty with allowing any other cultural or community belief system to express its own qualified reality. For instance, some may wish to say that Noah's Ark was old-testament real but without firm scientific evidence they should never say that it was classically real or scientifically real.

Obviously there are many belief systems with significant cultural value. Other belief systems are simply entertaining. For instance, astrological predictions entertain millions of people daily; some people seek out fortune-tellers and the readers of tarot cards, tealeaves and palms. These New-Age activities can arouse passions amongst scientists. Indeed many scientists find society's dalliance with such unscientific hocus-pocus activities deeply disturbing. However, these predictions fulfil a social role. Such forecasts can bolster confidence and encourage a waverer to undertake an endeavour or make a decision. Otherwise these ambiguous forecasts are not very useful, necessarily consistent or compatible with other competing forecasting systems. Admittedly what is disturbing is that a few people actually take such predictions seriously. There are people like this. I once knew a well-educated senior manager who apparently based all his important work decisions upon patterns of chicken entrails. It is a reflection on bureaucratic management that he was successful because, by chance, he made the best decision half the time.

However, we cannot be dogmatically dismissive. Indeed there could be a very small element of truth in some of these predictions. Recall that temperature is important in foetal development. Thus for instance, one might suspect that the time of day that we are born and the pre-natal and post-natal climatic conditions could marginally influence our personality traits and psychological development. Thus perhaps one's birth time and place might partially colour the way we see events happening around us. Nevertheless I must point out that currently all the detailed evidence is against this proposition.

How do I explain why am I being so accommodating to non-scientific viewpoints? Certainly it is not that I believe in relativism. All views are not equally worthy. Nevertheless superstition has long been a part of the human condition. Rightly we have never been content with *only* scientific knowledge. For aeons superstitions have filled the gap between the explanations that science delivers and our needs for a comprehensive satisfying framework of beliefs. However, I have faith that most people can distinguish the claims of science from those of religion, the paranormal and mythology. For instance, although modern science-based medicine is often too arrogant, few people believe that there are better alternatives for serious medical conditions. Nevertheless we must always endeavour to protect the young and intellectually challenged from misattributed truth. Anyway, this need for deeper explanations is why a modern belief system like that of eventual reality, needs to offer extra appreciations beyond the clinical realm of science. These appreciations must appeal to our inner spirits. Then, perhaps we can lay all superstitions to rest.

Science

Science and (to some extent) the broader discipline of natural philosophy (which embeds science) are marked apart by their particularly useful, expansive, consistent and enduring system of belief. This belief system is the engine of our modern society and western culture; it provides practical exploitable knowledge, adequate explanations of our surrounding, opportunities for identifying and solving problems, and rational answers to some deep philosophical questions. Thus it is a pity when decision makers ignore science.

Anyway let us see how we differentiate the rational and empirical picture provided by science from the beliefs of other areas of culture. We shall look at the question: What is science? The answer to this question is not as simple as you might think. Obviously science is that discipline on which scientists work. However, science relentlessly progresses. Thus scientists do not keep solving the same problems; instead they keep chipping away at the boundaries of scientific knowledge. If we had asked scientists say 100 years ago and 200 years ago to define science, then I expect that their answers would be quite different. Most likely the consensus answer would be different again today. However, I will not present any of these evolving answers.

I start from the privileged position of someone who understands some of the hypothesis of eventual reality, thus my answer here will be different again. Although I am maintaining that the expression of these historical answers would be materially different, these answers are interconnected. Each later answer effectively expands upon the earlier opinion. However, we are now able to give a definitive answer to the question – an answer that I believe does incorporate all previous impressions of the scope of science.

Inevitably we must start our explanation slowly. Note that science is like a multilayered onion. The disapprovers could say that science smells. Certainly one must always maintain a healthy scepticism about science; scientific theories do have their limitations. However, science's various strata are why I refer to an onion. Thus firstly I shall point to the historical progression of improving theoretical pictures, which provides one layered structure. Secondly I note that the scientific method is also an ordered sequence. The scientific method starts with observation, it regularizes these observations into empirical laws, it subsumes sets of laws into simpler scientific principles and finally it wraps consistent principles into a powerful descriptive paradigm.

Thus we note that science has a quadripartite structure; there are four essential distinctive components of science – just like the four seasons of the year. The first component is the body of experimental observation. This component holds the raw empirical data of science (eg observations on planetary motions). Independent experimenters will check these observations for objectivity add more data and attempt to improve the accuracy of the observations. This empirical data is analysed, the regularities are noted and general laws of nature are inferred from the data (eg Kepler's three laws of planetary motion). The set of all inferred laws of nature form the second component of science. One of the apparent mysteries of science is that the laws of nature seem not to vary from one place to another place, or from one time to another time. Hence this inference procedure works well. Now, experimental scientists will probe these laws seeking to find special cases in which the laws might fail. This is the process of falsification emphasized by the twentieth century philosopher Karl

Popper (1902-1994). Indeed Popper stressed that all natural laws must be capable of being overturned. Nevertheless it is only occasionally that any laws are shown to be need revision or even scrapping.

Other scientists investigate what lies behind these laws. These theoretical scientists attempt to build approximate models that represent the functioning of a small set of general principles, which subsume all the laws of nature. The set of all scientific principles form the third component of science. However, simple scientific laws cannot easily describe all empirical data. Thus these theoretical models are more likely to attempt to replicate the raw experimental data directly than the inferred laws of nature. We construct and refine models because of the difficulty of performing precise computations. Nevertheless the models represent the way in which we compare the principles with the empirical data. This checking may sound like a second example of falsification. However, we usually treat an apparent disagreement, between a set of well-established principles and an experimental observation, as a misinterpretation. There is no need to be pedantic.

Principles are so highly regarded that we look for any conceivable escape from the predicament before we are prepared to abandon a principle. This has proven to be the best approach. For example, in the early days of high-energy physics, physicists looking at cosmic-ray scattering images seemed to find evidence against the principle of conservation of energy and momentum. Rather than dismiss this principle, it was proposed that some unseen particle, the neutrino, was involved and that this elusive particle took away the missing energy and momentum. Subsequently this picture was confirmed.

Now, Friar William of Ockham (1288-1341) explained how we could choose the right principles. Many alternative sets of principles can explain the world that we see. For instance, we could propose a principle that there was a different sun in the sky for every day of the week, the apparent continuity of solar activity from day to day being an elaborate ruse. The slight effect (normally attributed to industrial pollution) that there is a correlation between the weather and day of the week, could be put forward as evidence supporting this extravagant suggestion.

However, Friar William proposed that when two or more principles or explanations (that equally fitted the facts) were offered, one should accept the explanation that is the simplest, the explanation without any extra restrictions or special cases. We call this procedure Ockham's razor and we regularly invoke this procedure to keep our picture of reality simple and sensible. Thus we do not introduce unnecessary principles such as the sun on a Monday is a different star from the sun on a Tuesday.

The final component of science acts as the keystone that locks the quadripartite components into a unified whole. This component is the set of beliefs providing the worldview. We have already described the worldviews of classical and quantum mechanics. The worldview is the picture and beliefs that underpin the principles of science. The worldview or paradigm attempts to make sense of the principles and subtly influences the development of principles and the discovery of laws. For instance, in the absence of evidence to the contrary we generally assume that the laws of science are the same everywhere in space. Thus no position in space is necessarily special from a scientific perspective and we do not regard the Earth as being the centre of the universe or even as being unique in terms of its life-bearing capabilities.

Laws are easy to test; a single contrary measurement creates an exception that will require either a modification to or replacement of the law. Theoretical principles are disprovable by the discovery of a new law that contradicts the principle. For instance, not so long ago, scientists thought that the universe had no preferred (left or right) handedness. Fate has decreed that most readers of this text will be right-handed – but this is presumably coincidental. Suppose we make a cellulose movie of a right-handed person having breakfast. Let us insert the film into the projector so that the projector illuminates the film from the wrong side. We will observe an image of a left-handed person eating cereal for breakfast. This parity-reversed world has the opposite handedness but otherwise, on the surface, it appears no different. Thus there was once a principle that the laws of physics were independent of parity (handedness). Consequently scientists were surprised when it was discovered that in special circumstances the laws of nature are biased – the matter of the universe is left-handed!

Now, beliefs are not easy to change. Thus a scientific worldview arrogantly resists change even more tightly than scientific principles. A rare observation that is not completely acceptable by the worldview is regarded as anomalous; it is termed exceptional and simply not believed. The successful prosecution of such an exceptional claim demands an exceptionally strong demonstration, way beyond the normal standard of demonstration. We are prepared to create elaborate excuses to explain the errors.

Thus the belief system is almost unchallengeable. Hence when predictions do occasionally fail, the paradigm sustains a flesh wound rather than a fatal blow. In the face of such intransigence, subterranean stresses accumulate. Eventually these stresses either subside or cause something resembling a cataclysmic volcanic eruption burying the old worldview under a new set of beliefs. Ultimately a paradigm succumbs to these stresses when a more successful new paradigm forces a confrontation and proves more appealing and successful. We often refer to this event as a paradigm shift; it involves the emergence of a new picture of reality. However, the paradigm shift is not instantaneous, the obscuring ash has to settle, the lava of new ideas has to stop flowing and solidify. Only then can we clearly see the full outline of new landscape. Note that this new landscape solidifies over the previous landscape. It therefore retains some old features.

Now, we have a way of choosing between otherwise equivalent worldviews. We use the notion of beauty. The mention of beauty may spark surprise in the ranks of non-scientists; however, scientists often talk about beautiful experiments, results, equations, ideas, etc. This is the same notion of beauty that is revealed in many works of art, but being scientists we also have a more technical idea about just what is meant by the term in the scientific context. Why are some things more appealing than other things? The answer is that we have an inherent archetypal concept of an ideal form for this thing. The closer the actual thing is to the corresponding archetype, the more beautiful it appears. In science the archetype is rationally founded – it is formed by logical argument. The archetype describes the ideal form. In art, the archetype is probably an emotional and experiential construct. Anyway if we cannot otherwise decide between worldviews, we go with rational beauty and reject the brute beast.

You may be surprised to learn that for almost a century we have had two dominant scientific worldviews. Furthermore, these worldviews are at odds; yet one has not displaced the other. Simply each worldview is supreme within a different realm of validity. These two paradigms share the materialist view that the universe contains space and matter. One worldview associated with the theory of general relativity successfully describes large-scale space. The other paradigm is associated with quantum mechanics. This paradigm successfully describes small-scale matter.

Unfortunately these two paradigms are incompatible in many ways. For instance, general relativity contains a term representing the energy density of space; this term provides for a force of cosmic repulsion. Cosmic repulsion explains a slight speeding up of the expansion of the universe that has (probably) been experimentally observed. We cannot calculate this repulsion term using general relativity but we can estimate it using quantum mechanics. Unfortunately the quantum mechanical answer is roughly one hundred and twenty orders of magnitude larger than observed. We would regard one order of magnitude as a severely worrisome problem. A disagreement of one hundred and twenty orders of magnitude is bewildering. It suggests that we have totally misunderstood the problem. Furthermore there are many other examples of problems encountered when we mix these two individually successful theories.

Scientists from both sides of the quantum-relativity fence have unsuccessfully tried to merge the theories. The problem is that separately quantum mechanics and general relativity have met many experimental challenges. Together these two theories spell trouble. I can now indicate the reason behind this incompatibility. It seems that the reason why we have not been able to merge successfully these two theoretical approaches is that we have not challenged the obvious presumption (common to both paradigms) that the fundamental universe is comprised of space and matter! This remark will become clearer later in the book.

Now, we are just over halfway through presenting our laborious definition of science. The qualities that we have identified so far do not uniquely identify science. In the time of the Greek philosophers, the emphasis of their natural philosophy was often directed towards logical and rational arguments used to define the archetypes. Furthermore these philosophers attempted to use these idealized forms to construct models, which directly interfaced with the natural world.

Hence we see that this quadripartite structure for science has been inherited from science's progenitor, natural philosophy. (I am resisting the temptation to call natural philosophy the mother of all sciences.) Natural philosophy had other less successful cousins. These cousins resemble the sciences. Thus conventional pseudosciences (like homeopathy and astrology) also style themselves on this same science-like quadripartite form. However, this does not mean that these pseudosciences quite succeed in meeting all requirements. Furthermore I shall annoy many when I say that extrasensory perception may become a proto-science (embryonic form that will emerge as a science) rather than a pseudoscience (a want-to-be science).

Recall that Popper proposed his falsification test as a method of differentiating science from such non-sciences. The reason why there are occasional arguments is that performing a falsification test can sometimes be too difficult to accomplish with sufficient authority to convince everyone. Nevertheless this quadripartite structure is still a key necessary characteristic of science. So before we continue with a complete definition of science, let us summarize in a table the details that we have just presented.

Table 1: Quadripartite form of natural philosophy and it sub-disciplines of metaphysics and the sciences.

Quadripartite components	Discriminated by…	Revealing
Experiments	Observation	Fact
Laws	Function	Type
Principles	Utility	Meaning
Worldview	Beauty	Truth

The worldview tells us what is real and what is true; experiments give us the facts. In between these extremes, the principles of science provide scientific explanations and the laws allow us to order and catalogue any real feature of the world.

Now, science has tended to ignore its role in asserting truth, preferring to point to its provision of facts. The founder of modern science, Newton, emphasized that he inferred his scientific principles entirely from factual observation. Newton was right to be sceptical of philosophizing. In his era, many rivals attempted to deduce logically what true reality must look like. These rivals had little chance of piecing all the parts of the puzzle together – indeed they failed. However, in creating the first consistent scientific picture of reality, Newton worked across all quadripartite components. Although he shunned the speculative nature of philosophy, he did not ignore the rational. Indeed to construct his theories of planetary motion, Newton made major contributions to mathematics such as the formulation of basic calculus – the branch of mathematics that deals with differentiation and integration.

Now, beliefs are those things that we hold to be true without conclusive evidence. For scientists these beliefs include the worldview picture, the methodology of science, mathematics and logic. Strangely we accept this worldview even though we know that science has not completed its quest to find an ultimate explanation of everything, even though we suspect that conventional science is only approximate, even though there are serious incompatibilities between some of our beliefs. This is the nature of scientific belief – a belief with an embedded doubt. We believe our worldview because it is the best picture that we have been able to construct to fit the facts of the universe. However, our acceptance is conditional. We can and very occasionally do completely change our minds (ie our beliefs).

This scientific growth represents evolution in action; hence the formulation of science resembles the struggles of life. There are occasional bursts of quickly changing viewpoints. Periods of slow growth, diverging opinions and relative stability of the conventional viewpoint separate these bursts of substantial revision. During the more stable eras new threats (laws) and capabilities (principles) slowly emerge. Each new paradigm has to cope with new circumstances but the paradigm inherits something of the previous requirements. The new paradigm must inherit the values of the old paradigm in some approximation. Thus when viewed in an appropriate way, the new paradigm reduces to its evolutionary ancestor. Consequently we build a progression of worldviews resembling a progression of ancestors, devolving back to a common ancestor. There are two conclusions I wish to draw from this discussion.

Firstly science is stratified; there is a sequence of pictures of reality with these worldviews stacked one upon another. Note that this layered format is also what we inferred after a discussion on the topic of Plato's cave.

Secondly science is something that evolved through the competition of ideas, concepts, arguments and hunches. Just as current anthropoid apes divide into several species: gibbon, orang-utan, gorilla, chimpanzee and human; current science divides into various disciplines: physics, chemistry, biology, etc. Each scientific discipline has its founder or founders. I acknowledge Newton as the founding father of physics and of science. Christopher Wren (1632-1723), astronomer and famous architect, apparently nominated William Gilbert (1544-1603) as the first practitioner of the 'new philosophy'. However, I am a theoretical physicist so obviously I am going to pick

Newton. Wren's 'new philosophy' was still very much an incomplete proto-science. For me, the worldview (the final piece) – not the use of the experimental method (the first piece) – is the most important. Science is quadripartite; science was incomplete before Newton.

Now, I need to make another aside before we can continue. The philosopher, Thomas Kuhn (1922-1996), who trained as a theoretical physicist, invented the term, paradigm shift. Strangely the concept of paradigm shifts has become a little contentious. Thus let us consider the issue. Paradigm shifts are rare. Hence most science is not concerned with paradigm shifts. Indeed there is an opinion that perhaps they do not occur. Some scientists liken science to a stream relentlessly heading downhill towards the sea. Usually the flow is slow and steady but from time to time the water runs over a cliff face and this sedate progress temporarily speeds up. This is an apt description of many phenomenal advances in science. Certainly the analogy applies to all advances arising from experimental discoveries. Yet this viewpoint misses the big theoretical advances that redefine the picture of reality. These pictures are important here, they describe our storeys of reality. Hence by highlighting changing worldviews, I am not devaluing the tireless contributions made by most scientists. I am simply looking at something else. Furthermore I do not support Kuhn's conclusions on where science could be heading. Kuhn suggested that evolutionary progress could be aimless – thus he missed the relevance of elegant beauty in science. Yet obviously I do support the idea that science evolves.

Of course *our* science evolved out of an ancestral Greek natural philosophy. Therefore we need one last discriminator to differentiate science from its metaphysical roots. Here metaphysics is that part of natural philosophy that is excluded from science.

Consistent with its etymological roots, we could attribute all of natural philosophy to science. Certainly the word, science, is derived from the Latin word *scientia* – the word apparently entered the English language (meaning knowledge) from old French in the fourteenth century. However, we ought to be discriminatory and not attribute all knowledge to science. Today's science is mathematical and there is a current belief (effectively modern materialism) that scientific reality is quantitative. This is not a new belief; Pythagoras held the same view. Thus in keeping with this deeply held conventional belief, I shall assert that (by definition) science deals only with the quantitative. This leaves any qualitative realm, which we shall discover does also exist, to the discipline of metaphysics. Hence this final condition completes our definition of science.

Consequently science is allocated the organized patterns of the universe whereas metaphysics describes the *orderless* characteristics of the universe. With this definition, a metaphysical state is without any measure of order; hence we should not call it disordered.

Thus let us recapitulate what we have decided. Science is an evolved descendant of Greek natural philosophy. Science has a quadripartite structure comprising observation, law, principle and worldview. Science is the study of the quantitative nature of reality. Science builds a sequence of pictures of reality. The higher pictures ignore the finer detail of lower pictures. These finer details are irrelevant to understanding the more macroscopic behaviours. This is the nature of the magnificent achievement known as science. Hence this is our completed definition of science and it is summarized in the table.

Table 2: Science, through all ages, is defined as the discipline that satisfies four simple requirements.

Factor	Requirement…
Ancestry	Descendent of Greek natural philosophy
Classification	Quadripartite components (see previous table)
Explains	The quantitative
Evolution	Series of stratified realities and worldviews

If we take away the quantitative requirement, we call the larger discipline natural philosophy. Furthermore the difference between natural philosophy and science is called metaphysics. Note that these are acceptable definitions. They coincide with standard usage of such terms. They incorporate older meanings and they will stand up to future developments. Thus these definitions will assist our thinking when we investigate what lies beneath doctrines such as materialism.

Universe

I shall use the word, universe, to mean absolutely everything that exists without exception. Furthermore science is the arbitrator on disputes about what exists. This definition of universe applies whatever form the universe takes ie the definition does not imply that there is necessarily any space, any matter, any galaxies or any intelligences. Now, it may be an empirical fact that the universe exists, but we still want to know why the universe exists. An acceptable answer will need to be a rational answer ie the answer needs to be logical. Yet can we divorce logic from the universe in which it applies? Is it acceptable to use a logical argument to explain the existence of the universe? Surely logic cannot exist independently of a universe. We really need to justify both the universe and logic together.

When posed in this way, this problem appears insurmountable. How do we even start to solve the problem, if we cannot take the existence of logic as a given assumption? Hence the Greek philosophers argued that logic was self-evident and therefore logic did not need justification. Likewise these philosophers believed space was the embodiment of a self-evident Euclidean geometry. Today we know that physical space is only approximately Euclidean. General relativity teaches that gravitational fields curve space-time and render space-time non-Euclidean. Likewise classical logic is not the only expressible form of logic. Thus we cannot treat classical logic as self-evident either. Nevertheless, in these discussions, we will continue to use classical logic until we demonstrate the validity of logic.

Let us consider a conundrum about logical arguments. This conundrum goes straight to the heart of the problem. We are going to be parochial; after all, we have expectations about our universe. Thus we shall presume that there are some valid *ab initio* logical arguments that someone can understand. Now, we very much want this presumption to be true. Nevertheless we do not always get what we wish for. On inspection, there are two necessary aspects to this presumption. Firstly there must be some valid logical arguments that are true. Secondly these arguments must have a finite number of steps.

Generally the steps in an argument are proven by invoking other beliefs. Ultimately this repeated reduction will not yield an acceptable *ab initio* logical solution unless we invoke a genuine tautology or equivalently a tautological loop. In logic, a tautology is the only thing that needs no justification. A tautology is something that must be true precisely because it is offering no new information! For example if the number two is defined as the sum of one plus one, then the statement 1+1=2 offers no new information.

However, the idea of basing the notion of reality upon a tautology is extremely perplexing. Tautologies are self-justifying but they present themselves as insipid statements. How could we use a tautology to justify anything significant, let alone such fundamental conceptions as logic, the universe and everything? (Let me apologize to the late Douglas Adams for this deliberate corruption.)

Luckily nothing is ever quite what it seems. We shall discover that it is precisely because we have to justify logic along with the universe, that opportunistic freedom reigns. We have already mentioned that we expect a multi-storey conceptual edifice will represent reality. Every storey of this edifice of reality will be quite different and we will forge this hierarchical structure to create science and logical descriptions. In this way, what we wish for can come true.

Number

Few normal people stop to ask the elementary question: What are numbers? However, we have chosen to identify science by its quantitative character. Thus we will find it very useful to consider the nature of the numbers we regularly use. From this discussion we can ascertain the general concept of number. This description of the general number concept is conventional. Nevertheless a clear notion of number is important – please try to follow these arguments.

Surprisingly it took a long time for the full understanding of number to evolve. Thus this dialogue will follow a logical rather than a strict historical order. To start this exposition I need to explain from where everyday numbers come. Of course we all have an intuitive grasp of some numbers eg the natural numbers: 1, 2, 3, 4... These are useful numbers but this sequence of integers is incomplete. The natural numbers only include the positive-definite integers. Here 'positive-definite' is mathematician's jargon for definitely positive ie it does not include the number zero which cannot decide whether it is positive or negative.

Now, all numbers come with rules that say what mathematical operations we can perform on these numbers. We may only use those arithmetic rules that are appropriate for the type of numbers in question. Thus let us take any two of these natural numbers and add them together. We will get another natural number as the result. This is one of the basis rules for integers. However, it would be useful to have an 'undo' rule allowing us to reverse an addition. This undo rule defines the subtraction operation.

Unfortunately if we subtract one positive-definite integer from another positive-definite integer the result may not be amongst the positive-definite integers (eg 3-5). In retrospect it is easy to ensure that the rules can always provide an answer, we have to add further integers: 0, -1, -2, -3… We owe this development to Hindu mathematicians in the seventh century. In this way we devise an integer arithmetic that makes good sense, it will always generate answers using the + and – operations.

Thus I can add three oranges and four apples to get seven fruit. However, we would like to be able to use an algebra to ask what-if questions. I want to share equally my three oranges with a friend. I ask: What number added to itself makes three? It is disappointing that our integer arithmetic does not have an answer. We can accept that there is no answer, or we can expand our concept of number.

By adding an unknown number to itself, we are accomplishing an integer multiplication (repeat addition). So what happens if we include the multiplication operator in our algebra? Can we form numbers that work with both addition and multiplication operations? Clearly if we multiply any two integers, then the answer is an integer. However, as before, we also want an undo operator, called division. If we divide an integer by another, we do not always get an integer. Hence we resort to introducing further numbers, like 7¾ that we can generate as the ratios of finite integers. Mathematicians felt that these numbers were sensible and hence they called these numbers the rational numbers. In this way we have formed an arithmetic based upon addition and multiplication. Actually there is a unique problem with division by zero but why let one small number interfere with the big picture?

Arithmetic is not everything. We still would like to be able to use an algebra ie to solve brainteaser questions involving multiplication. For instance I have one more orange than I have friends. Each friend has the same number of oranges as I have. Together we have three oranges. How many oranges do I have? If you have not followed the convoluted associations, do not worry. That is why they call such problems brainteasers. Algebraically the question becomes: What number can I multiply by itself to give three? The answer is strange. Clearly some of my friends cannot be all here!

Well, no sensible rational number solves this puzzle. This unwanted fact is not particularly recent; indeed results such as this greatly disturbed Pythagoras. Fearfully the Pythagorean school of philosophers resolved to hide the secret failings of the rational system of numbers from the public. Perhaps we should not be too worried about a weird problem with oranges and strange friends. However, suppose we have a cube with length, width and height all of one metre. Pythagoras' theorem tells us that the separation of two diametrically opposed corners of the cube is the solution of the same problem. Surely the length of this diagonal should be expressible as a number? After all, we can measure the distance with a ruler.

This embarrassing conclusion brought to a halt the ancient Greek's serious study of arithmetic. Unintentionally their own endeavours had undermined the status of all philosophers. Mathematics was advertised as an eternal truth; thus mathematics was thought to provide glimpses of the divine. Scholars with an understanding of mathematics were to be revered; they were the people closest to the immortal. Not surprisingly the Greek philosophers sought to ignore algebra and subsequently focused their mathematical vigour on geometry. In today's status-ridden society there are lesson for all of us here. Everyone is special; nobody and no group should be so vain as to claim any general superiority over others.

How can we alleviate the problem that worried the ancient Greeks? Well, we proceed as before and further generalize the numbers. By repeating the same unknown number in a multiplication, we are accomplishing a mathematical process known as integer exponentiation (repeat multiplication). This repetition is more familiar when we refer to an unknown number squared. Squaring means raised to the power two. Our earlier queries ask what unknown number when squared will equal three. Thus the next expansion of the number concept occurs when we seek an arithmetic permitting explicit expressions of powers of numbers.

However, if we raise a rational number to a rational power the result is sometimes rational but often it is not rational. The answer to our nominated problem is known as the square root of three and written as $3^{1/2} = \sqrt{3}$. We cannot compute such expressions unless we allow a new category of numbers that are like $\sqrt{3}$. These numbers are called irrational because they cannot be written as *sensible* finite fractions. We may approximate $\sqrt{3}$ to any degree of accuracy by a rational number expressed as a decimal number described by a finite number of digits eg the familiar $\sqrt{3} \approx 1.732$ that you see in four-figure tables of square roots.

However, the true value of $\sqrt{3}$ is only expressible by an infinite number of decimal digits. This decimal sequence of digits cannot even end with a finite-length repeating string of digits (like $4/33 = 0.121212...$). Hence there are no shorthand ways of writing the number as a decimal. This indefinite-length decimal behaviour is the hallmark of an irrational number. Thus take any irrational number and add any finite string of digits anywhere in the decimal sequence and you must get another irrational number. Wow, there is so much scope for these irrational numbers that, not surprisingly, they greatly outnumber the rational numbers! Irrational numbers are not rare isolated anomalies. The rational numbers are the rarities. Now, the complete set of numbers made from the rational numbers and the irrational numbers is known as the set of real numbers. However, there is more to come!

We accomplish the undo operation for exponentiation by raising the resulting number to the reciprocal of the original power. The reciprocal of the power is obtained by dividing one by the power, for example $\frac{1}{2}$ and 2 are reciprocals, $\frac{5}{8}$ and $\frac{8}{5}$ are reciprocals. Thus the inverse operation is formally the same exponentiation operation. Nevertheless you should not be surprised that there is a still a difficulty – in fact there are two major issues.

As a repeat multiplication, we can evaluate expressions involving negative numbers raised to integer powers. However, a negative number raised to a non-integral power is very different. For instance, the square root of minus three, $\sqrt{-3}$, does not exist as a real number since the square of all real numbers is always non-negative. Once again, we are faced with a choice. We can accept that some equations have no answers or we can expand our number system so that we can ensure that there are answers.

Indeed I am sure that you have guessed we shall find it profitable to expand the number system. Usually we accomplish the expansion by introducing the so-called imaginary number, $i = \sqrt{-1}$. Any real-number multiple of i is called an imaginary number. We call the combined system of real and imaginary numbers the complex-number system. This system is portrayed in a new manner. Recall that we can place irrational numbers next to rational numbers on a line of number (eg a ruler). Well, the real and imaginary numbers cannot sit alongside each other along a line. Thus we place the imaginary part of a complex number on a different axis of the two-dimension space of complex numbers. This two dimensional space is often called the complex plane or Argand plane.

The first major issue that we face is one of non-uniqueness of some mathematical expressions. A (complex) number raised to a real-integer power has a unique complex value. A (complex) number raised to a real-rational (non-integral) power has a finite set of different solutions. A (complex) number raised to a real-irrational power (or a complex power) has an infinite set of different solutions, all equally valid. Consequently algebraic expressions involving a (complex) number raised to a complex power can prove quite challenging.

Let me give an example for the mathematically inclined. Do not be concerned if you cannot follow the detail. I intend this example to be a reminder for those people who once knew, but have conveniently forgotten, how dangerous exponentiation can be. We start with the recognizable principal-value expression for the exponential function

$$\exp(2\pi ni) \times 1 = 1$$

where n is any integer. We raise both sides of this equation to the power i and simplify the powers

$$(\exp(2\pi ni))^i \times 1^i = \exp(-2\pi n) \times 1^i = 1^i$$

or

$$\exp(-2\pi n) = 1^i / 1^i .$$

It is easy to be confused by the significance of this result. Here 1^i is actually an infinite suit of solutions generated by multiplying or dividing by the irrational number, $\exp(2\pi)$.

$$1_n^i = \{\exp(-2\pi n)\}.$$

The second major issue is perhaps more surprising. Most likely, you learnt some complex arithmetic at secondary school. However, what your teachers did not tell you was that there is an alternative number system, which can equally well handle the square root of minus one. We should not necessarily use complex numbers! This alternative number system was developed by the Irish mathematician, William Hamilton (1805-1865). Hamilton's quaternion mathematics introduces three new distinguishable numbers that are each the square root of minus one. Hence quaternion mathematics generates a four-dimensional space of numbers but otherwise this algebra is quite similar to that of the complex numbers.

Some readers may be familiar with complex numbers but unfamiliar with quaternions. Quaternions are an important part of the hypothesis of eventual reality. Thus let me make some introductory remarks about quaternions. We can think of a quaternion as a doubly complex number. Let us start with the complex number written as $z = x + y\hat{j}$. Now, suppose that we require that both x and y also behave like complex numbers but with a second separate imaginary number $\hat{i}$. Then, we write

$$x = a + b\hat{i} \qquad y = c + d\hat{i} \,.$$

$\hat{i}$ and $\hat{j}$ designate the numeric terms providing imaginary-style extensions to a real algebra. Thus naturally we make their product a third imaginary-style number ie $\hat{i}\hat{j} = \hat{k}$. Hence our doubly complex number becomes

$$z = a + b\hat{i} + c\hat{j} + d\hat{k} \,.$$

This result is a quaternion. Consistency requires the three imaginary-style numbers anticommute. The quaternion has a scalar part, a, and a 3-vector part (b,c,d) with (anticommuting) vector components each acting like the conventional imaginary number.

Now, we have directed this discourse towards showing how we explain everyday numbers. Hence indirectly we have revealed what all numbers (not just those we have discussed) really are like. Numbers and the rules that they obey go hand in hand. More technically we can say that numbers are the elements of an algebra. An algebra is a set of rules for manipulating its elements ie the numbers. Each specifies the other.

Thus there can be many types of numbers. Different types of numbers obey different sets of rules (known as axioms). We have been interested in rules for addition and multiplication. Note that I have not given these special rules – most people intuitively know them.

However, I discussed these two mathematical operations and some of their associated numbers. These two operations ($+, \times$) have proven useful in solving practical problems. Yet this empirical observation does not imply that the discussed types of numbers are going to be relevant for our discussions of fundamental reality. We will have to discover the algebraic operations with which any numbers are to work. Then, this knowledge provides the basis for choosing which type of number is fundamentally useful. In this regard we note that complex numbers and quaternion numbers are the elements of the only two (integer-generalized) algebrae that will allow us to solve general algebraic equations using mathematical operations $+$ and $\times$.

Do you now see numbers in a new light? Almost everyone has rather preconceived ideas about the naturalness of common numbers. Yet when we look at what a general number is, we see that a lot of specific information goes into the system of numbers that we regularly use. The set of numbers does not have to include the integers 1, 2 and 3. Numbers do not have to obey rules of addition or multiplication. Thus we have a lot of everyday mathematics in need of an explanation. We cannot take the familiar everyday numbers for granted.

3. Fundamental Reality

In the previous chapter, we laid the necessary foundations for our analysis. We explicitly defined four interrelated terms: reality, science, the universe and number. It is rare to see these terms defined. Indeed I suspect that many people might explain these core terms in subtly different ways. Therefore please be appreciative of any differences. Remember, in this document, words mean what the author intended.

Thus before we proceed, let me summarise these four definitions. Reality is the sum total of everything real, as interpreted by a single worldview. There are many worldviews of natural philosophy and science. These worldviews form a multistorey hierarchy that we shall call the edifice of reality. This edifice defines all consistent views of the universe. Thus the universe is everything real, as interpreted by *all* the worldviews of natural philosophy and science. Here science is the numeric part of natural philosophy. Numbers are defined as the elements of an algebra; an algebra is a finite set of consistent rules (the axioms) describing how to perform binary operations. (In this context, binary refers to an operation on two numbers yielding a numeric result.)

Hence we have completed the introductory preparations. Thus we can now start our analysis, which is to be based upon these definitions.

Methodology

Therefore let us continue our discussions within the framework of generally recognized logic. How else could I write this book? Actually we will be able to justify this approach soon. However first, we must discover how to make a start to our deliberations. Where do we begin an explanation for the necessary existence of the universe? Perhaps we should start by seeing why common methods of solution pose difficulties.

Well, most rational arguments are composed of a finite sequence of steps starting at some natural beginning. These linear arguments begin with a set of given assumptions and the arguments build to their inevitable conclusions. Hence these conclusions logically follow from the starting assumptions. At first glance, we seem to be at a disadvantage. Where can we begin? We would like some imposed starting requirement from which we can launch an argument to explain the universe. Yet we have no external system of anything to create the universe. (Deists please note that the universe denotes everything that is mystical and otherwise.) Undoubtedly we have to reconcile ourselves to the fact that there is no possibility for any externally imposed precondition on the universe.

However, this just-stated omission is itself a necessary precondition. Hence more appropriately we state that there are no imposable *a priori* assumptions about the universe, other than this denial of any further preconditions of any kind. We shall refer to this statement as the no-input requirement. This precondition will become our revised definition of the universe. There is nothing lying beyond the universe that can shape it, create it or give it any specific properties.

When we work with logical arguments, some results can seem trivial and hence unimportant. Other arguments may be difficult to follow and we might suppose the conclusions are more important. However, this is not true. Of course all valid results are both trivial and important. Obviously the no-input condition is trivial – it says nothing that is unexpected. However, the no-input condition is important because of the way it says nothing unexpected!

Thus consider the starting proposition that the universe of everything is everything. This is essentially our definition of the universe. We are not referring to the physical cosmos of space and time but to everything: material, mental, spiritual, conceptual, etc. This tautological expression appears quite uninformative. A universe of everything must be the totality of everything! However, we can add spice to the proposition by acknowledging the meaning of the word, everything. Thus we more meaningfully say that there is nothing other than the universe. Note this statement does not imply that there is a universe. We do not want to slip in any assumptions. Clearly there is only one practical implication of this statement – there is nothing else having any affect upon a universe (if it were to exist). This is our starting premise – called the no-input precondition – that we are about to solve.

Therefore we have found a natural beginning for constructing the logical argument that could explain everything. Awkwardly the no-input condition is an implicit rather than an explicit requirement. This requirement does not explicitly state any of the details of the forcing of features or behaviours that we are to exclude. Hence we are going to need an unusual way of extracting the implications of this lone implicit precondition.

Now, let me make a short digression to reveal the size of the problem faced. There are many self-supporting reasons why we know that the universe must exist. You, the reader, must exist. After all, that which does not exist cannot read a book. If you exist, then the universe exists. Wow! How significant does that make you? However, we need to turn away from such Descartes-style empirical evidence offering no explanations. Instead we shall look for an entirely rational argument. We want a logical argument that can explain the universe.

Nevertheless just any valid argument may not be enough. It will take a clever rational argument to validate existence. For instance consider an example. Specifications of the conditions of both existence and non-existence generate knowable information. Furthermore the definiteness of any information demonstrates the logical existence of at least some information. Existence is a demonstrative property that is attributed to something else. Many years ago the Greek philosopher, Parmenides, argued that non-existence is definable only as the lack of existence – thus it lacks the direct significance that we can attribute to existence. Hence non-existence is only possible if there is an existence to which it can be related.

You might wonder whether, with this Parmenidian-style argument, we have confirmed that a universe of something must exist. Unfortunately there is a huge self-imposed flaw in the reasoning. We may not presume that logic is applicable to the universe – this imposition is contrary to our no-input precondition. Thus we have become a victim of our own cleverness, hoisted with our own petard! This example shows the sophistication of the problems that we face. We have to justify the use of logic, and explain the universe! Too many people needlessly despair at this point, and decide to choose an easier problem.

Naturally some researchers make progress by supposedly rendering the problem easier to solve. They try to form an answer by burdening the problem with their prejudices. There are plenty of obvious encumbrances. For instance they could add the precepts of scientific materialism, pre-space monads, abstract logically related structures, Berkeleian mentality or semantic linguistic forms.

Furthermore these researchers are free to add any refinements. Thus they can force the resulting hypotheses to conform to observable reality. Hence we are left to judge a difficult beauty contest: Is the short hairy Neanderthal more attractive than the noisy giant Gibbon or the playful miniature Bonobo? (By the way, the noisy giant Gibbon has previously ripped the arms off dozens of unimpressed judges.) Well, if there ever is to be an answer, if we are ever to live up to the illustrious title, *animans scientifica*, we must abandon all prejudices. Inventing assumptions can only make our problem more difficult. We must tackle the no-input problem head-on.

Fortunately the no-input precondition opens another door providing a rare opportunity for using a particular approach. This mode of attack involves forming a self-consistent solution to the otherwise impossible-to-solve quandary. Indeed this method should be familiar to many scientists. Overwhelmingly complex demands do not intimidate scientists. The natural world is a rich exhibition of amazingly complicated phenomena. Thus even though the principles of science are simple, sophisticated complicated-looking tasks riddle science. Hence scientists have adapted many techniques to tease solutions out of impossible-looking problems.

Consequently the standard procedure for tackling an implicit problem involves seeking a self-consistent solution. Often this method includes the iterative convergence of a trial solution. However, we have enough clues about reality to avoid adding the iterative process. Thus we shall base our solution methodology purely on the self-consistent verification. Nevertheless iteration is available in principle if we were confused enough to make a mistake. I say confused because, as supposedly smart citizens of the universe, we should already know a lot about the answer.

What does this approach entail? Well, I shall not describe the general method, just the way we shall use it to solve the current problem. We shall fashion three steps in the method's prescription:

1) We make a few appropriate weak assumptions.

2) We use these assumptions with no-input condition to derive a potential description of the universe.
3) We show that this derived universe explains and validates all the weak assumptions of step one.

Now, how does this method work? It works by a circularity of justification. Thus the appropriate weak assumptions (of step 1) will cover what we need to generate a solution. This set of assumptions includes the applicability of some form of logic and the existence of something. However, we will not say what form of logic nor say what this universal something is like. Then, in step 3, we effectively self-justify the introduced weak assumptions. In other words, the circularity of the procedure demonstrates the self-consistency of the solution. Thus we show that the weak assumptions are part of the solution not the method. Hence the technique demonstrates that we will have generated an adequate solution for a universe subjected to the no-input requirement. Indeed we know (from the no-input requirement) that no other available preconditions can rule out a description generated by this procedure! Simple, isn't it? The weak assumptions generate the solution and the solution justifies the weak assumptions. Let me make it clear. We are not saying that the solution only satisfies the weak assumptions; the solution must justify the weak assumptions. If you are still puzzled, it may be because you cannot see how a solution can *justify* the weak assumptions. Bear with me until you see the specifics. It will become clear in retrospect.

Anyway, we will not know that the obtained solution is unique. Nevertheless we may identify our solution as a self-authenticating logical sequitur of the no-input precondition. Significantly our result is also the only solution of a type that satisfies the chosen minimal assumptions. Furthermore we are, in effect, testing these minimal assumptions by confirming whether the solution is consistent with the no-input precondition. Hence I refer to this three-step procedure as the *by inspection* method.

Let me digress again for a moment. I want to discuss tautologies. In logic, we define a tautology as a statement that can only be evaluated as true. Thus it is seen to be trivially correct. For example, Tom is the biological parent of either an odd number or an even number of human children. The statement would be true, even if Tom were my pet chrysanthemum. Likewise 1+1+1=3 is a tautology when we accept definitions like 2=1+1 and 3=2+1. Note that we need various definitions in order to understand the tautological statement. These definitions answer questions such as: Who is Tom? What is a parent? What is a child? Now, all definitions are true by default. Thus these appropriate definitions and the statement form a tautological set of statements that are each individually true. However, the validity of the tautological conclusion (eg a mathematical relation) may not always be immediately clear (eg $\sqrt{1{,}522{,}756} = 1234$). (Just think, all those arithmetical problems you solved as a primary-school pupil were just dumb tautologies!)

Likewise in the *by inspection* method, we are discovering the nature of the universe by using a tautology. Our initial impositions comprising the no-input requirement and the step-one assumptions are our definitions. The universe deduced in step two is just a reinterpretation of these definitions. Nevertheless we have initially solved the wrong problem. We want a solution of the no-input requirement. If we have extra requirements we have not accomplished this aim – we have definitely not solved the no-input problem. Hence we need to remove the privileged status of definition from the additional assumptions of step one. This is the reason why in step three, we ask that the solution (for the universe) authenticates the weak assumptions of step one. With this authentication completed, we can justifiably claim to have found a solution of the no-input problem. Thus without any further input assumptions, we have created a tautological self-authenticating loop. Hence this loop is somewhat analogous to the mythical self-creating tail-devouring serpent, Uroborus, of ancient Egypt. Our self-authenticating logical loop will provide justification for the existence of both the universe, logic and hence itself. These statements of existence (of universe and logic) are to be members of our set of weak assumptions! Additionally the exposed solution will tell us what the universe and logic must look like.

Assumptions

We have intimated what some of the weak presumptions of step 1 are to be. However, we have not yet spelt these assumptions out. Next, we will explicitly specify three assumptions; although, in some sense, there is only one actual composite assumption. We shall see that each assumption is eminently reasonable.

Firstly we assume that a universe does exist. Of course this assumption is implicit in the other two assumptions about the universe. Nevertheless it is appropriate to state the existence requirement as one of the explicit assumptions that fix the character of our necessary solution. Then, our solution will show how this assumption plays an important role.

Secondly we assume that the universe will display a space. We will not demand a specific form (eg a particular dimensionality or geometry). We certainly will not equate the universe with space and its contents. We simply ask that some aspect of the universe looks like a geometric space. There are two reasons why this assumption is included. Clearly we want our solution to hold a science that we can compare with known science. However, more importantly although it may not be obvious *a priori*, our third assumption requires that there be a space.

The third assumption is that the universe be rational and logical. You will be aware that within the context of the traditional worldview there is no clear relation between logic and space. We will find otherwise; but this is an *a posteriori* result. Hence it is useful to state explicitly this spatial assumption also. Ultimately the requirement – that the universe has a rational logic to which it abides – subsumes our three assumptions. Now, there are two related aspects to this assumption of rationality. First the universe must exhibit explainable rational behaviours, and second a form of logic must appear as a natural ingredient of the universe.

These are our set of weak assumptions. Could we have chosen different assumptions? Clearly this method of attack does not preclude selecting different presumptions. In principle, depending on the nature of the new presumptions, we might easily obtain the same solution. Alternatively we might obtain starkly different solutions, which do not admit the veracity of the three chosen presumptions. This is the reason why we select presumptions that are as non-restraining as possible whilst permitting us to generate a useful solution. By useful I essentially mean sensible. It could be possible that there are other non-trivial solutions to the no-input requirement. Such solutions are not particularly pertinent. We are not interested in other solutions, which will not concur with rational explanations. Indeed if we could show that any other solutions exist, then perhaps we should take these solutions to be irrational reinterpretations of the rational universe. After all, it is easy to invent a many-to-one non-sensible correspondence with a rational universe. Perhaps I should also point out that the *by inspection* method could not be used with an irrational solution. Thus we are probably safe assuming that the solution is unique.

Now, there are several considerations relevant to choosing the set of weak assumptions. Clearly we should keep the number of assumptions to a minimum. This will make it easier to confirm that a solution is fully self-consistent. Furthermore we should not make any assumptions that could be contingent on choice or chance. For instance, we ought not to assume that you will eat our favourite dinner tonight. A deduced solution can only legitimately authenticate universal necessities. Contingent details, like the name of this book, whilst being incontrovertibly a fact of our universe, did not have to be so. Assuming such contingent facts would be in direct contravention of the no-input requirement.

This leads to the question: How do we determine what is necessary and what is contingent? In practice, we may use experiences from our universe as a guide. However, in principle, we can only accomplish this classification after we have deduced a solution. Thus the identification of necessity is also the role of step three of the method. Step 3 confirms that the assumptions are all necessary. Thus we shall be stating that our solution, obtained using the *by inspection* method, is a tautological necessity.

For those readers finding the going hard let me try to simplify how we are solving the problem. We solve the no-input problem using the *by inspection* technique. For a minute, think of a different problem, say, what number when added to itself makes four? We could try the number two as a solution. Thus we find that the number two satisfies the problem in the example. Then, we argue that the answer two is a solution to the problem. Of course there may be other solutions that we have not tried. Well, we do something similar in step 3 of our methodology. We try a solution to see if it fulfils the requirements of the problem. Where do we get this trial solution? In our methodology we create the trial solution in step 2 using the weak assumptions of step 1.

Now, the *by inspection* technique does not tell us whether there are other solutions. This is why we have step 1. Step 1 tells us how we are to identify which of all the possible solutions we want. The minimum requirement that we have for a useful solution is that the universe is comprehensible. This is the heart of step 1 – we say that the solution must be rational and potentially comprehensible to somebody. Thus we see what is actually meant by weak assumptions – a discriminator to identify acceptable solutions. Then, we add the no-input requirement and get the trial solution (step 2). Finally we perform the more elaborate task of showing that the solution satisfies the no-input problem outright. Hence this is our three-step procedure for solving the ultimate no-input problem. It sounds easy. Furthermore despite having fooled everyone for so long, it is surprisingly easy.

Foundation Apeiron

By choosing the weak assumptions, we have completed step 1. Thus let us make a start at solving step 2. We shall simply accept the weak assumptions that the universe exists and that the universe is explainable. At this stage in our search to discover the nature of fundamental reality, we will not need to use any other assumptions. However, we will need the other weak assumptions to complete the solution's methodology and thereby verify the acceptability of the result. Thus our immediate goal is to uncover the appearance of fundamental reality.

Now, we shall follow convention and name this ultimate foundation of all reality, the apeiron. The early Greek philosopher, Anaximander (611-547 BC), coined this name. The Greeks were prolific neologists (creators of new words and meanings). Hence there is a common saying *the Greeks had a word for everything*. In one sense, this saying must be false. Clearly the Greeks did not have a specific word for items of modern technology eg the television set's remote control. Yet the ancient Greeks did invent words for a great many concepts, and they did have a word for 'everything' – namely the apeiron. This apeiron is the founding substance of the universe. I use the term, substance, in the sense that we can construct everything from the apeiron.

Anaximander correctly surmised that we may not impose external properties upon the apeiron. Hence every characteristic of the apeiron must be self-generated from within the apeiron. How must the apeiron appear if there is no regularity, scale or shape imposed from outside? If nothing makes or holds the fundamental universe in any particular form, how will the apeiron self-assemble? We shall soon confirm the remarkably simple answer. The universe adopts a bland indefinite form. In principle, no 'part' of the universe can be functionally different from any other 'part'. This is why Anaximander argued that there could be no boundaries and hence the apeiron was of infinite extent with no breaks. Essentially we will not find systemic features without a necessary reason.

Naturally some scientists may worry that a well-known process called spontaneous symmetry breaking could self-form order in the apeiron. The freezing of water provides a typical example of what we mean by this process. The haphazard arrangement of molecules in a liquid can become regular in a crystal and thereby randomly select preferred crystalline directions. However, to get such symmetry breaking we have to impose externally (eg via the scientific laws) a choice of options upon the system. Thus we can rule out spontaneous symmetry breaking in the apeiron. The apeiron is without external interference; it is the ultimate anarchic freedom-land. There are no objective laws, directions or scales. It is orderless.

Of course to the uninitiated this description of an indefinite apeiron sounds far-fetched. Surely we could never construct anything resembling the regular scientific world from the seeds of this freedom-land? Well, do not let prejudice blind you to the facts. As the builders of the leaning tower of Pisa realized too late, considerations of structures should be secondary to those of a firm foundation. Let us get the foundation correct before we worry about the links to recognizable reality.

Thus we conclude that the apeiron does not sit within a space-time framework; it has no ancillary features such as matter or thought; it is without any notion of boundaries or borders and it is even missing the conceptual elements and laws of logic, mathematics and science. How can we describe that which is so featureless? The answer is we describe the apeiron in a bland manner. After all, this is the very essence of a modern tautology. Tautologies are not meant to add substantive information.

We know that the nature of the apeiron is not reliant upon any external impositions. However, we are assuming that the apeiron does adhere to an intelligible description. Hence, in order that we can describe the apeiron, we will assert that a description – involving one or more generic or archetypal forms – can represent the apeiron. What we are really saying is that we shall build a model representation of the apeiron. This model portrays the view of the apeiron that we have from the vantage point of our experienced rational world. The apeiron describes the ground storey of the edifice of reality – the foundation of reality. As our Greek forbears noted the apeiron does not resemble our empirical world. Hence inevitably the empirical world must sit on one or more higher storeys of a multi-storey edifice of reality. Now, this model strategy provides a useful approach to solving the puzzles of reality. Clearly we need to describe reality in effable terms ie in terms that we can discuss.

What are we saying? We are accepting that the universe must exist and the universe is made of something. However, what do we mean by this phrase? Is this something its parts? Individual constituent parts cannot actually have a potential existence that is independent of a whole universe. If we only had one part, then we would still have a whole universe. Thus the notion of a part of the whole fundamental universe needs to be clarified. When we refer to parts of the universe we are actually referring to identifiable bits that are described as consistent forms. Importantly these parts must relate to one another. Indeed every part must have a relationship to at least one other part (or a system of parts). This is what it means to be a part. Any part that had no relationship of any kind with anything else would be an independent solution for the universe in its own right. Such an isolate would not be a part of something else.

In our directly experienced world, examples of such consistent forms are ideal representations (eg a perfect circle, an indestructible atom) that nature never quite realizes in practice. However, it is reasonable to presume that nature does not precisely display such ideal representations because these forms are non-basic constructs. We should not be surprised if Plato was right; the ultimate fundamental elements of the universe exhibit the simplicity of perfection.

Before we discuss the form of these parts, let us mention the status of fundamental science. Current so-called fundamental scientific theories have proffered a variety of elemental forms eg point particles, one-dimensional strings or higher-dimensional branes. Suggestions like these are unsuitable for the foundation of reality. We are not allowed to assign any specific quantities to the fundamental elements. Scientific elemental forms possess values such as dimensionality, size, energy and spin. A fundamental form cannot have any imposed dimensionality; a fundamental form cannot have any inherent qualities. Thus similarly the fundamental form cannot be a continuum – we cannot attribute qualities or quantities to be distributed over a dimensional space. Consequently forget such entities and think bland.

Clearly if there are no inherent qualities, then we cannot distinguish one fundamental form from another ie there can be only one type of fundamental form. We call such a fundamental single style of elemental form, a monad. Now, two questions confront us: How many instances of monads are there? How do monads build a universe exhibiting qualities and quantities?

Let us consider the first question. How many instances of monads are there in the apeiron? Obviously an explanation would be needed for imposing any definite number of monads on the universe. We could perhaps accept there might be a special reason for there being no monads or just one monad. However, we will not waste time looking for such a special reason. These examples would make no sense. The absence of any monads would mean no universe – a weak assumption has ruled this possibility out of current considerations. One solitary monad is no better. A featureless monad is a non-entity. A one-monad model would permit no possibility for creating qualities and hence no rational model – contrary to our assertion that there is an explanation of everything.

Inevitably we are unwilling to assign any other non-special number (eg forty two) without a clearly imposed (non-fictional) reason. Hence we are left with only one possibility. We conclude that there must be an indefinite number of monads. This lack of an imposed limitation is really just a consequence of the no-input requirement. There are no limits placed upon the number of monads. Thus we must have an infinite repetition of this fundamental unit of the universe.

Answering the second question lets us complete the picture of the apeiron. How do monads build a universe exhibiting qualities and quantities? Our monads are discrete units with no intrinsic qualities. Any information that is attributable to the monads may only come from some associations between monads. However, the apeiron is non-spatial; we cannot attribute distances or directions between monads. Consequently we are solely left with a non-quantitative pointer functionality to provide a mechanism holding the entire descriptive content of every features of the universe. This pointing is the minimally elemental function from which we can deliver any rational judgements and forms.

Next we shall introduce a new terminology. We shall refer to monads that are pointers (with no other characteristics) as events. These events are so-named after the hypothesis located on the edge of science, the hypothesis of eventual reality. Apeironic events are not to be confused with any events that might happen in your household everyday. Apeironic events do not sit in space and time; they are simply pointers. Such events only point to other events. This linked unending network of events is our rational model of the apeiron.

How should we view this argument? Well we have asserted that we must not posit any special features for the apeironic foundation save that it is understandable. This is the heart of the argument. We cannot ask for any space or spin, rich or red, mind or matter, friend or frog, but we do ask for existence of logical forms and relations. These are the conceptual features needed to deliver an understandable apeiron. Hence we argue that the logical forms are monads – all of which experience true equality in the apeiron (because we cannot make them different). The logical relationship is the reference to other monads (because they have no properties). We call such pointing monads, events. We can picture this system of pointing as a network. However, the network and its events do not exist in any space or at any time.

Clearly we need to do more to complete the *by inspection* proof. However, the picture that we have just derived is all there is to the fundamental universe. This is what ultimate reality looks like. Do not be dismayed by this picture. Obviously the elementary universe is necessarily quite elemental in its representation. Furthermore this solution was derived by using a tautological loop.

Hence we declare that the fundamental universe is an unlimited qualitative network called the apeiron. There is nothing difficult to understand in this picture. Thus perhaps the more interesting questions are how this apeiron builds the amazing richness of our experienced world with space, matter, conscious minds, logic and science. However, please do be patient; we are coming to these issues.

Now we get to glimpse the reason for all of existence. Within the context of this model, non-existence of the universe is just not an option. Non-existence would require that every event refuses to point to any other event. However, there are an infinite number of events. We would need to outlaw pointing in order to stop every event from manifesting existence. However, we cannot impose any input constraint to stop events from pointing. Hence we can be categorical. The chance that every single unimpeded event would not point to any event is guaranteed to be exactly zero! Thus the model generates a self-consistent certainty that the apeironic network must exist. This is the crucial argument that authenticates the proof. The other self-consistent features will latch onto this justification of the universe's existence. Hence the apeironic model solution requires that the universe must exist. Thus what we have to decide next is whether the model has all the requisite features to be a valid solution of our no-input problem.

Please do not be alarmed at my casual use of words – like refuse, decide and choose – to describe the action of events. Such word usage entails a little poetic license. Unfortunately there are no dedicated words to describe suitably the behaviour of events. Do not imagine that events are debating issues and making rational decisions. However, no other events or systems of events have any say in the pointing behaviour of an event. These events only point, and this pointing is uninfluenced by anything.

Note that this apeironic picture is trivially fascinating; we present the apeiron's nature and existence as both obvious and necessary. The no-input requirement ensures that these events are overabundant and carefree. They create the universe because we cannot stop them from existing, they point because we cannot stop them from pointing. The pointing of the lawless events knows no external constraints. Thus the possibilities generated within the apeironic network are endless and unlimited. Freedom-loving opportunity is at the heart of fundamental reality.

First Principle

In the last section we formed the groundwork for a solution to the no-input problem. However, we still have to perform the lengthier step 3 of the *by inspection* method. How do we proceed? Well, as Plato appreciated, we invoke the first principle. Plato recognized that there was a need for a first principle of the universe. The first principle describes how the fundamental substance forms the higher-storey characteristics of the universe and hence how we come to understand the universe.

Recall that the founder of Greek philosophy, Thales, believed the fundamental substance was water. Nature provides water with attributes that allow it to exist in solid, liquid, gaseous and plasma states. Thus the very earliest first principle was that everything in our world is a blend of these four states. Thales simple empirical picture made a lot of sense for his time. All the liquids Thales knew were aqueous. He saw how water would evaporate into the air from rock pools. He saw how cool air could form into clouds, which can then drip rain. He saw how water could freeze to form solid ice. It was an interesting hypothesis. Now we are searching for something similar: some principle that will explain how the universe has its current form and yet does not impose further attributes or assumptions upon the foundation apeiron.

So what is this principle? Well, we have no external source to generate the principle. Consequently the principle can be little more than a tautology in its own right. Naturally we are regarding tautologies as good. However, we cannot prove tautologies wrong. Hence sometimes people like to disparage tautologies and call them unscientific. This attack is a blinkered misunderstanding of science. Science is quadripartite. Thus a tautological law could be a useful rational mnemonic, whereas a tautological principle must be of prime significance and a tautological belief must be pre-eminent.

Now, today's physics has many powerful principles. Almost without exception, these principles' roots and justifications are empirical in origin. We promote these simple and elegant principles as the progenitors of the natural laws of all the sciences. Thus these principles embody the common themes that are evident within sets of laws of nature. Consequently the laws become seen as an expression of the combined action of all applicable principles to a particular perceived situation. Furthermore there is a status hierarchy of principles, with more powerful principles expressing the concepts, which pop up in several more-restricted principles. Thus as we go up the hierarchy there are fewer principles and the principles become more general and abstract. Unfortunately the physics of the twentieth century left us with a fair number of principles at the top of the heap including the following principles: causality, determinism, continuity, uniformity, rationality, effable explanations, analyticity, materialism, a mathematical universe, relativistic covariance, etc.

I am being somewhat arbitrary in listing so few principles. Note that I have demoted principles related to the behaviour of matter. Many physicists might have started with the conservation laws (energy, momentum, spin, charge, lepton number, hadronic colour, etc.). Perhaps they would have added Newton's third law (action and reaction), the second law of thermodynamics (entropy non-decrease) and some of the principles of quantum mechanics (Heisenberg's uncertainty, Pauli's exclusion, state evolution and measurement). Then, there are the favourites of the high-energy physics community such as CPT invariance and the theorems of Goldstone and Noether. It is surprising just how many major principles there are.

The hope has always been that we could crown the hierarchy of principles with a single logical principle from which all other principles followed. Indeed we will discover that this is what the hypothesis of eventual reality has achieved. Eventual reality elevates a pre-existing physical principle. I did not even mention it in my above lists. This principle was born out of physical study but we now recognize it as a statement of mathematical logic – it has a status befitting the first principle.

This ultimate principle is an updated version of Fermat's principle and it plays the role of Heraclitus' *logos*. The principle dictates how underlying processes rationally shape our perceived world. For the most part, the apparently orderless commotion that takes advantage of the opportunities of the apeiron is self-defeating – but there is the potential for assembling order even here. This is the first role for the first principle. Thus this first principle instructs us how almost miraculously to fabricate order and consistency. Furthermore the first principle also shows us how to build all the elevated storeys that make up the edifice of reality.

Now, you have all heard the Darwinian slogan: *survival of the fittest*. In this generic form, the slogan is a tautology and part of the belief system of science. However, we also consider that this Darwinian phrase is also a principle because we conceptualize the meaning of the phrase as it applies to evolution in the natural world. For instance, we can consider the gene pool of a population. Genes that confer a reproductive benefit to an individual or close relative are more likely to prosper at the expense of genes conferring a lesser benefit. Such selective evolution is taking place all the time with different common drivers such as diseases, diet, life style and appearance. Consider again the Black Death, which struck Europe in a series of waves principally from the fourteenth to sixteenth centuries. Of course there are many reasons why the first wave had the most devastating effect, reducing the population by one third. Certainly societies developed better precautions after each successive wave. However, the smaller consequences of later waves were also due, in part, to changes in the human gene pool. It is only logical to accept that any genes, which could increase an individual's survival from the plague, would have become relatively more common amongst the survivors of each successive wave.

Now, there is an equivalent tautological belief in eventual reality. For historical reasons, I will refer to this conceptualized idea as Fermat's principle. It appears that Pierre de Fermat was very good at spotting important relations but perhaps he was not always quite as successful at completing the work. In mathematics, the theorem that became known as Fermat's last theorem was proposed in 1637 and remained unproven until 1994. However, I am assuming that Fermat did not have a miraculous proof that was too long to fit on the margin of the page. You have probably heard the intriguing story, which I shall not repeat.

Now, physicists were a little quicker with the unsolved problems Fermat set them. For instance, to understand Fermat's optical principle of least time, we only had to wait a few years. This principle needed the advent of the wave theory of light as formulated by Christiaan Huygens (1625-1695). However, it did take a little longer than it should have. The wave theory of light had a serious competitor and thus Huygens' wave theory took quite awhile to be generally accepted.

Now, I shall apply Fermat's principle in a more general form that I shall state evocatively as: domination by the dominant. Clearly this statement is a tautology. Hence, it needs no further justification. This is why the principle is so important. The principle is of course just like Darwin's principle. Thus Darwin's principle should be viewed as a special case of the first principle.

I need to explain the technical ins and outs of the principle. Therefore using familiar situations let me give some gedanken examples of how we might apply this first principle. These examples will help to describe all the important issues related to this principle.

Imagine a large crowd of people who are all participating in independent conversations on their mobile phones. From a balloon hovering above the crowd – you cannot make out any particular conversation – all you can hear is the general hum of voices. However, ten percent of the crowd are New Zealanders ('Kiwis') who have secretly plotted to start singing their national anthem at a prearranged time. The clock strikes eleven and the singers begin. Even with only a scattering of New Zealanders in the crowd, the hum of the other voices is irrelevant, the Kiwi voices dominate. We hear their collective song only because they are singing in unison, only because their voices complement each other. Whoops! This is not quite what happens. New Zealand has two national anthems. Half of the Kiwis sing *God Save the Queen* and the other half of the Kiwis sing *God Defend New Zealand*. Nevertheless the hum is vanquished and we hear two competing songs that are confusingly intermingled and superimposed.

To understand what we hear, we look at the correlations between individual voices. Before eleven o'clock there were no correlations. After eleven o'clock there were two subsets of correlated voices to drown out the incoherent hum of the other voices. Fermat's principle lets us identify these correlations and assign a strength value (the number of singers) to their overall resulting voice. In a similar way, Fermat's principle lets us to examine the apeiron and construct the dominant coherent resultant of that undisciplined network of pointers.

Let me give another everyday example. Suppose you happen to like to own and use old technical equipment. Then, naturally you have a pendulum clock on your wall and an outstandingly versatile still camera that captures images on a cellulose film! You set up the camera to irregularly take ten thousand snaps of the clock without winding the film on between snaps. The camera takes all the snaps at a very fast shutter speed and the film is very slow. Thus the superimposed image on the film is correctly exposed. What do you expect to see? You will not see a total blur; neither will you see the pendulum in the average position at the bottom of the swing.

The resultant image shows two partially blurred pendulums, one at each end of the swing. Otherwise there is no sign of a pendulum between these extremes. The explanation is simple. In the middle of the swing the pendulum is moving very quickly, it does not stay around this region very long and consequently the pendulum is not snapped here very often. Thus the cumulative film exposure with the pendulum at this position is too low to register. In contrast, at the extremes of the swing the pendulum is momentarily stationary, close to the extremes it is moving very slowly. When the pendulum reaches either end of a swing, the pendulum hangs around these spots relatively longer and is therefore much more likely to be snapped at the extremes. The camera's film is adding up all the contributions. Contributions from the same place in the swing are in supportive agreement. The first principle expresses the mathematics detailing how strong each resultant contribution is going to be. The stationary extremes win; they signify what the outcome of the image looks like – two pendulums.

Now, I shall not present any detailed mathematics but this will not prevent me writing some iconic equations. All powerful theories have simple sound-byte equations. Most adults will recognize the scientific equations $E = mc^2$ and $F = ma$, even if they are not quite sure what the equations represent. (I am somewhat unhappy at abusing Newton's legacy by writing the second equation in this popular manner. Newton actually correctly said that force was the rate of change of momentum ie $F = dp/dt$.)

In fact iconic equations such as these display quite limited characteristic information about their corresponding theories. Likewise eventual reality has a fundamental characteristic equation that represents Fermat's theorem. This equation, the equation upon which all of quantitative science is based, is written $\delta S = 0$. We use this equation to identify the harmonious agreement existent within a discordant commotion. The summation of the dominant coherent voices dominates this harmony. The dominant configuration is said to be mathematically stationary; this is written $\delta S = 0$.

In the annals of science, we have already noted that this expression is a most powerful equation. For instance the equation represents a starting point from which, in a classical context, we can derive the equations of general relativity and quantum mechanics. Of course the classical picture of reality is not fundamental and the two starting classical pictures (used to derive these theories) are different. Thus in the hypothesis of eventual reality, we shall look to our more fundamental picture of a reality (the apeironic network) to found our theoretical principles.

Now, there are other notable aspects to the particular way that we shall use Fermat's principle. Recall that scientific descriptions of nature are stratified. These alternative pictures of science correspond to different viewing scales. For instance consider a pure liquid in a jar. A large-scale picture says this liquid is a continuous fluid characterized by its pressure, mass density and temperature. However, the small-scale picture describes this liquid as a chaotic assembly of discrete molecules with a distribution of velocities. On an even larger scale view, the jar of liquid is just part of planet Earth. On an even smaller scale view, we describe the motion of the subatomic constituents of the molecules. As we change our viewing scale, we talk about different issues. We are going to use Fermat's principle to generate this hierarchical sequence of pictures.

It is no accident that we always get a change of scale. This change of scale provides the scope for our summation of contributing terms. We derive the details of each macroscopic picture from the summation of dominating microscopic contributions. The scale change generally ensures that all details of the microscopic picture are lost in the macroscopic representation. Hence the macroscopic interpretation is free to be based upon new concepts.

Thus each time we generate a higher storey to the mighty edifice of reality we shall expand the viewing scale. By adopting this much coarser view, we repetitively create a differentiation between a microscopic and a macroscopic descriptive realm. Only with this scale change can the natural interpretations of these realms be quite different. Hence we get an expression for the higher-storey description of the large-scale detail from the resultant sum of very many lower-storey microscopic contributions. Inevitably there are so many contributions that, in general, most independent microscopic contributions are unlikely to build a significant macroscopic structure. Only particular sets of dominating terms build an inevitable coherence to reveal the macroscopic structure of the higher storey.

Now, there is something special about these dominating microscopic terms. Obviously these dominating microscopic terms add supportively. Let us consider what this implies. Well, either the contributions are effectively equivalent or some part of them is equivalent. Consider an example to help explain the relevance of this equivalence. Suppose we make a series of small waves in a long straight stretch of a canal. If we do this at random the surface of the canal becomes a little choppy. However, suppose we can build each new wave to sit upon the earlier waves. Then instead of a ruffled surface we get one large wave travelling down the canal. This creates a dominant behaviour and it occurs when the contributing waves are in phase. Thus in this example the phase parts of the dominating microscopic terms are all equivalent.

We can attribute a reason for this equivalence. The reason is expressed in terms of the concepts of the macroscopic picture. We say that the equivalence arose from a (macroscopic) symmetry. For example, if there is a left-right symmetry in our body, then our left hand and our right hand should be the same size. However, strictly the causal attribution should be the other way around. Because our hands are the same size there is a left-right symmetry. We can express this viewpoint more generally. The dominating sets of microscopic terms (assigned by the stationariness condition) generate a symmetry that is displayed on a *small* scale of the *macroscopic* picture.

Now, the resulting layers of science are special in another way. When we create a macroscopic picture of scientific reality we deliberately ignore the small-scale happenings belonging to the microscopic regime. Thus there is no clear remnant of such microscopic irregularities in the resulting mathematics of the macroscopic formalism. We can understand this imposed ignorance as a process by which we smooth over all small-scale behaviours. Consequently by expunging this finer structure we generate a new picture that has all the hallmarks of the mathematical property known as analyticity. Analyticity is a critical feature of all scientific formalisms. It is illuminating to realize that analyticity is simply the way in which we extrapolate over the microscopic features that reveal the lower storey of reality.

This excising of small-scale detail in the macroscopic picture could worry some thoughtful readers. Thus let me discuss this concern by explaining how it differs from a misleading example. Think of the classical orbit of an asteroid around the sun. Suppose there is a small error in its known position and velocity. Then, the difference between the predicted orbit and actual orbit will progressively grow over time. Thus small-scale input errors produce large-scale errors in prediction after long periods of time. Well it is correct that, in the same way, our macroscopic picture cannot account for the microscopic detail. However, this omission is more a feature than an error. By design, we choose the macroscopic picture with new stable descriptors. Then, we correctly incorporate the statistics of the large number of microscopic contributions. Hence there is no manifestation of this omission in the macroscopic variables. Consequently the details of individual contributions are irrelevant.

Hence this is what I mean by Fermat's principle. Fermat's principle is a self-authenticating principle. This tautological principle says that the dominant do dominate. We will use this principle to generate all of science with its stratified multiple descriptions of nature. The stationary condition identifies the dominant contributions. Each application of Fermat's principle leads to a change in scale, the possibility of descriptions based upon new concepts, and macroscopic pictures displaying symmetry and analyticity.

Thus this principle is critical to the hypothesis of eventual reality; this is the first principle of the universe. The first principle is the creator of our science and the rationality of the universe. We can summarize it best by listing the lessons revealed by the examples: The largest sets of coherent contributions are dominant. We identify the dominating sets of coherent contributions by the stationary condition. We create a new macroscopic picture of reality by summing microscopic contributions and looking for the new elements of stability. In the resulting macroscopic picture, we interpret the microscopic coherence as arising from a local macroscopic symmetry. By overlooking the irrelevant microscopic detail, we create an analytic macroscopic representation.

The first principle and the apeiron are the source of the wonders of our universe. Intriguingly the concept of an apeiron has been around for ages. Many philosophers and scientists have debated alternative formats for the apeiron. Just imagine how the world could be today, if any historical philosopher had appreciated the role of this simple first principle. Maybe they could have constructed the sciences from pure rational thinking centuries ago. Perhaps we could have had an advanced technological society long before the tide of humanity so dominated the terrestrial natural world.

4. First-Elevated Storey

We are well underway on our quest for the ultimate answers. At the foundation of reality, the apeiron is a model of limitless possibilities. We have uncovered the tool that will let us build the strata of reality. Hence we may now start our climb to these higher realms.

Scale

Let us proceed cautiously. We should not overlook several important early steps. This first topic of interest has to do with measurement scales and why we need them. All numbers have a corresponding scale, even if that scale is an abstract discrete unit. Recall the cube of unit side length we mentioned earlier. The 3-D diagonal across diametrically opposed corners has length $\sqrt{3}$ units. Clearly, for a physical cube, this length is measured in units displayable on a real ruler. Conceivably we could change the units on the ruler and make the diagonal length 3 units. However, we cannot remove the irrational number requirement. The side lengths are now $\sqrt{3}$ units. Anyway we call such a change of units rescaling. The Pythagorean equation used to derive this diagonal result cannot depend upon the scale marked on our ruler. Thus Pythagoras' equation is scale invariant ie the result still holds when we scale all distances by the same factor. Inevitably all our quantitative relations are similarly relative. Hence the purpose of introducing a scale is to remove an endless chain of comparisons. Then, we can make an assertion relative to an agreed common standard. This is the benefit of a reference scale.

However, the apeiron is not quantitative; it has no given size or scale; the events have no meaningful pointing direction or pointing distance. Nevertheless to deliver quantitative science and to uncover a space we must attribute a scale to the universe. The scale will provide the reference unit against which we can compare all quantities. Clearly there is only one possibility for such a reference scale. Thus we attribute the scale to the event's sole function. Hence we assign a fixed unit separation length to every pointing reference. For instance consider any pair of events with one event pointing to the other event. Then, we deem that these events are one universal unit apart. This universal unit will prove to be the basis for every other unit of measurement.

Are you wondering if this unit roughly corresponds to the Planck distance scale? Well we cannot assume that it does. We need to prove everything. However, yes – it does. Clearly both measures represent the ultimate scale limit of any spatial distance. Yet we have a long road to travel to get to this correspondence. The connection shows up eventually via the scale for measuring the curvature of space within a derivation of general relativity on a classical storey of reality. The Planck distance scale is staggeringly small. This distance approximates to $1.6\ 10^{-35}$ m. Expressing this value in a different way, a million million million million million million (ie traditional British sextillion or US undecillion) Planck distance units would be just 16 metres.

Note that we attribute a unit length for numerical counting purposes. We do not arbitrarily impose a direction on any of the network links. However, we will try to assemble the network into an arrangement of the smallest possible dimension.

Now, if the network had only a few events, then we could seat the network in a quantitative space. To picture this spatial network, imagine you are playing with a set of balls and rods, traditionally used for building models of molecules in chemistry. We suppose the rods are of fixed length and are allowed to stick into the balls at any angle. We can easily make a network with just three or four balls. However, we cannot make a spatial network for the entire apeiron. The apeiron has an infinite number of events; inevitably the model's minimum number of necessary dimensions grows with the number of events.

An infinite network will require an infinite number of dimensions for the arrangement that the network defines. To make matters even worse, there are an infinite number of ways to make an infinite network with infinite dimensions. Thus by imposing a scale we find that we have not placed the network in a space. This is because, as we shall soon see, a space must be finitely dimensioned. Nevertheless by forming quantitative arranged structures we have made an important start.

Thus we have made our first step. We have added a scale upon which we will be able to base all measures. Indeed now that we have a scale, we can apply Fermat's principle to this structure. This is how we will be able to ascertain the dominant coherent macroscopic form of the dimensioned apeiron. This dominant form will be our space.

Space

Thus we are almost ready to show how we generate a space, the foundation space, from the muddled network that is the apeiron. However, we have yet to explain what constitutes a space. Indeed space is so ubiquitous that it is easy to overlook the need to explain what space is. I am using the term, space, in a general sense. I am not specifically referring to that three-dimensional expansive structure in which matter is located nor specifically referring to the relativistic four-dimensional continuum with coordinates taken from this three-dimensional space and from time. These are both examples of spaces, and these examples do show the essential features that characterize all spaces.

However, we cannot arbitrarily impose upon the universe any empirical style for space. In particular we should not assume that space is the habitat for matter and energy fields. We want to discover what form of space is rationally formed from the apeiron. Thus we need to decide what we mean, in general, by the concept of space. What is the characteristic that cries out *I am space*? Well let us think about what is critically necessary for a space.

Firstly we recognize aspects that ensure that space is certain. Thus space is a mathematical construct. Space is a continuum with a fixed finite number of coordinate dimensions. This is another way of saying that space is quantitative and part of the scientific world. Thus we are implying that space is analytical and that we will derive its style by applying the first principle. This application is why space will be continuous. We are going to expunge the microscopic network from the new macroscopic formalism of space.

Furthermore we are insisting that we describe space by a fixed set of coordinate variables. This number of variables defines the finite dimensionality of space. Note that we would be misinterpreting the concept of dimensionality if we identified a variable number of dimensions. In such a case, we would be required to introduce further dimensions to parameterize any apparent changes in dimensionality.

Furthermore we cannot have an indefinite (infinite) number of dimensions. Infinities are often worrying in mathematics and they are disastrous for a count of dimensionality. We want to allow a notion of how far apart points in space are. Thus we are going to have to add terms together. For a finite dimensionality we will add a finite number of terms. Everyone should remember that the order in which we add a finite number of terms is irrelevant. Hence the summation 1+2+3 agrees with the re-ordered summation 2+1+3. Unfortunately the same equivalence is not clear for an infinite number of terms. Think of a summation like +1+(-1) +1+(-1) +1+(-1)... and various ways of re-ordering the sequence eg (-1) +(-1) +1+1+(-1) +(-1) +1+1... or (-1)+(-1)+(-1)+... + 1+1+1... . Therefore we cannot even place clear bounds on the results of general infinite summations.

Thus an arbitrary infinite series is potentially senseless. Clearly, in general, we may not stop after any finite number of terms. Hence we can never reach the end of the sequence of summed terms. This is why we require a definite, finite number of dimensions for a describable space.

Secondly we require properties that underscore the specific functional role of space. We shall insist that space recognizes translations and rotations. A translation simply means that we can move from one point to another point in this space. A rotation means that we can spin around. These operations go straight to the essence of what is meant by a space. A finite translation always takes us somewhere else. In contrast a finite rotation could take us back to the starting viewpoint. A rotation rules out of contention a one-dimensional space. This is reasonable; the purpose of coordinates is to allow us to specify a relation between their components. This one-dimensional coordinate would describe a no-where point simply varying in a coordinate we are likely to call 'time'. We cannot allow such a one-dimensional representation to be called a space.

Now, some readers may be familiar with the way science usually regards space. They will note that, as usual, I am being more general and less restrictive than is normal. Historically the Greeks viewed space as the logical embodiment of Euclidean geometry ie they imposed definite rules of geometry. Let us forget that many philosophers also had further strange ideas eg fundamental space was an unchangeable plenum. Now, Newton inherited and modified the geometrical suggestion. He saw space as an independent quantitative entity. His space was a fixed continuous background in which matter was located and matter was moved by well-defined forces according to given laws. The absolute nature of space worried Newton but he had to accept it. Eventually Einstein replaced Newton's independent absolute space with a continuous relational manifold interdependent upon matter. Effectively Einstein let space be rubbery – space could be distorted and stressed. Gravity, as caused by matter, is one form of distortion. Permitting this distortion changes the form of the geometry. Thus this is how we normally describe everyday space. However, the description is too specific because it presumes that matter and energy pervade all spaces. This description also requires a specification of dimensionality. We will discover that such an everyday space is not the foundation space. Thus we take a less constraining definition.

Therefore let us summarize our technical explanation of the term 'space'. We define space as a definite mathematical continuum structure of finite dimensionality, and with rotation and translation operations.

Foundation Space

Next, we will show how a finitely dimensioned spatial structure can emerge from the localized coherence of the apeironic network. As we form this space, we will also manufacture the associated mathematics. Mathematics is born from our assumption of a logic, and our consequential ability to count sets of events. Counting provides the integers. Then, from this system of integers, we build a mathematical number system.

Now, we are treating space as the working blueprint for an organized quantitative coordinate system. We look for a suitable coherent blueprint using Fermat's principle. This suitable coherent blueprint must agglomerate the coherent attributes of the apeiron, and possess localized mathematical translation and rotation symmetries. In our first act, we increase the scale at which we can resolve features. Hence we will deliver a macroscopic continuum. Note that there is no need to demonstrate that there are not other ways to form the spatial representations. All we need to do is show that there is a solution with the requisite rotational and translational symmetries.

I hope that you can recall the various relations embedded in Fermat's principle. Fermat's principle requires the resultant coherent stationary space will exhibit symmetries. Consequently we ask for a solution of Fermat's principle, in which these symmetries are the translation and rotation symmetries required of a space. The mathematical structure that describes a symmetry is known technically as a group. A group is a specialized algebra in that a group is just an appropriate set of rules with group elements that satisfy the rules of the group. These rules describe the functioning of the group operation that mirrors the symmetry. I shall be presumptuous and represent the group operations for translation and rotation by symbols $+$ and $\times$ respectively with corresponding group elements called numbers. Clearly these elements are some type of number as they are the elements of an algebraic set of rules. However, we have yet to discover whether the $+$ and $\times$ operations are the familiar addition and multiplication.

Let me give the simple explanation first and then the more technical explanation for those who wish to follow it. I am using the symbols $+$ and $\times$ for obvious reasons. Translation along a line is an operation looking like addition, just as rotation about an axis resembles a multiplication. If you are familiar with the complex plane you might recognize the rotation operation as the action of multiplying by $\exp(i\theta)$. A rotation (about the origin) corresponds to multiplication by a number of unit modulus (ie unit length). This is a way of saying that rotation does not change the size of a number just the orientation of the number.

We can almost guess the form of mathematics we need for our fundamental space. Scientific mathematics is based upon the complex numbers. Complex numbers obey a set of rules known as the integral-domain axioms. However, we are being less specific than this, we are asking only that the solution conforms to the translation, $+$, and rotation, $\times$, group operators. In practice, this means that we can also allow quaternion numbers, which satisfy a slightly less restrictive set of rules. Mathematicians call these less restrictive rules the division-ring axioms. The relaxation in axioms is tantamount to recognizing that the order in which we make out-of-plane rotations is important.

It is obvious that we need to relax this axiom (about the order of rotations). Think of two orthogonal 90° rotations of a car in three dimensions and you will see that this is precisely what we empirically expect. Position a car on a road facing north. Then, look down on the car and rotate the car 90° anti-clockwise so that it faces west. Now rotate the car 90° clockwise about a north-south axis. The car will be pointing straight up in the air. However, suppose we perform these operations in the reverse order. The first operation flips the car on its side. The second operation turns the car to face west. This is a different result. Hence because we cannot interchange the order of the operations, we say that these rotation operations do not commute. Now, by dropping the commutation requirement of the integral domain we obtain the axioms of a division ring. Next, we will examine how this requirement arises in a little more technical detail.

In this derivation, numbers and space go hand in hand. We source a measurement scale from the microscopic count of events. Thus along a line of numbers out from the origin we identify the translation operation + with addition. Think of a number as a line from the origin out to an endpoint. Then, translation of the endpoint will almost certainly change the length of the line (ie magnitude of the number). Furthermore rotation of the endpoint about the origin will almost certainly change the direction of the line. In general, rotations and translations do not commute. (We cannot interchange the order in which they are carried out.)

Consequently we have to form a component representation for the spatial coordinate of the endpoint. The general coordinate is described as a summation (+) of components which are formed as the product (×) of the separate number elements (magnitude and direction) of the two symmetry groups. Here the direction-providing group elements have unit magnitude and the magnitude-providing group elements are without direction. When we add coordinates we add the magnitudes corresponding to the same component directions (as defined by the rotation group). When we multiply coordinate components we individually multiply their directions and multiply their magnitudes. Hence we must also identify an appropriate rule for the product of magnitudes.

We start by observing that rotations obey a relation known as the distributive law. Rotate a short line through 90° and the magnitude will not change. Now split the line into two arbitrary parts. This time rotate each part through 90° and then add their lengths. The results should be the same. Hence the product group obeys the distributive axiom. The distributive axiom relates multiplication and addition. It implies that we can uniformly stretch a short line and its length is still the sum of its stretched parts. This is the axiom that requires that the real numbers rather than rational numbers pack the line of numbers.

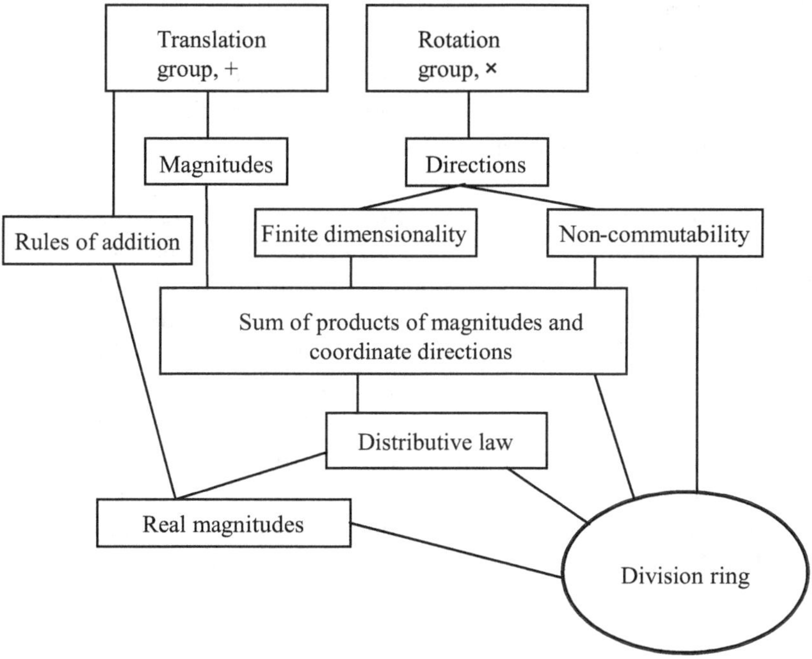

Figure 2: The figure summarizes the tangled web of inter-relationships that leads to the identification of numeric magnitudes with real numbers and the dual translation and rotation symmetry with division-ring numbers (ie real-number axioms without multiplicational commutability).

If we only required the summation (+) operation along a line, then we would only need rational numbers (ie rescaled integers). However, we are also asking for rotation and so our axioms have two algebraic group operations. It is strangely cyclic. The properties of numbers define their algebraic axioms and the algebraic axioms define the numbers. Thus we are sharing another (semi-)group operator along the line of numbers. This operator obeys the distributive law. Hence we identify this second operator, obeying the group axioms, with the familiar product of magnitudes. In turn this leads to the requirement for real numbers defining the algebra populating the line of numbers.

We need to treat the origin as a special case since as the line shrinks to zero we do not have a finite-magnitude inverse on the line. Thus the two group symmetries with the distributive law provide the specification for the algebra of a stationary space: a division ring over the real-number field.

To summarize, the translation and rotation groups, the line of real numbers and the so-called distributive law provide the information that is at our disposal. These requirements define the local algebra of space as that of a mathematical division ring.

Now, there are two finitely dimensioned algebraic division rings: those of the complex numbers and the quaternion numbers. As both of these solutions satisfy the symmetry criteria, our solution for a space with rotation and translation will expressed both. After all, there are no conditions, which can restrict space to either possibility alone. Thus the formal solution needs to be a combination of both solutions. Before I forget, let me remind you of an important fact. These symmetries are only expressed accurately in the small-scale limit of the macroscopic representation.

However, there is a further technical problem combining complex numbers and quaternions – the simple product of these algebrae is not itself a division ring. Thus in order to allow nature to be expressed by both of these local symmetries, we must provide two groups that are genuinely distinct. We can make our local spatial coordinates of this foundation space the product of a complex number and a quaternion number. However, each of these algebrae must have its own distinct set of mathematical signs (±).

Furthermore there are more complications. There are twin parity solutions (eg mirror images) for both algebrae. Each alternate-parity form is equally stationary (dominant) and hence each form will emerge as part of the coherent macroscopic representation of the underlying apeironic milieu. We refer to these different parity forms as branches. Note that complex numbers describe two dimensions; quaternions describe a set of four dimensions. Together these division-ring algebrae describe the foundation of an eight-dimensional (4×2) space. The set of branches expands this space. To appreciate the branches consider reflections in a mirror. Recall that a scene and its (parity-inverted) reflection in a mirror fuse on the surface of the mirror. In one of his novels Lewis Carroll imagined that Alice could walk through the looking-glass mirror. In a similar way the different space branches meet across what are known as branch cuts. So far, we have mentioned four (2×2) such branches; as we shall see soon, there are actually eight.

We have not quite finished with the apeiron. Those apeironic events, which dominate in the stationary contributions of sets that describe the spatial branches, may be associated with places in the derived macroscopic space. However, these events (that are placed in space) may have additional links to unplaced events. These unplaced links form the basis for a qualitative form known as the aethereal field or simply the aether.

I have been slipping the verbal term *field* into the text. It is a word with many meanings. Therefore let me clarify how I am applying one of these meanings. You will probably be familiar with phrases such as gravitational field and magnetic field. In this context *field* refers to something described by quantities with numerical values distributed throughout space. The (macroscopic) average density of matter throughout the universe would be another field. On the other hand, the aether is different. The aether is non-numeric and is not really a part of space. Nevertheless the aether is that part of the macroscopic apeiron that is seen from space. This association, with locations in space, makes the aether a weird non-numeric field.

We reference the aether from a macroscopic space that has been formed using Fermat's principle. Thus the aether is also a smudged coherent remnant of aspects of the apeiron. Clearly this referencing is subjected to the same averaging as the general scaled apeiron. Furthermore we are restricted to accessing the aether from within space. Thus the aether becomes subjected to space's symmetry conditions. If we step forward and then step back, or turn completely around the aether is unchanged. Consequently, whilst the aether is not quantitative, the aether inherits the symmetries of the complex and quaternion group operators. These operators describe the aether's own internal dimensions as the reversed reflections of the convoluted dimensions of eight-dimensional space.

Now, the handedness of the aether provides the final parity label for distinguishing the branches of space. Once again each parity form is independent. Hence we have derived a picture of an eight-dimensional space with a set of eight (2×2×2) branches. This is the description of the foundation space. A space is a natural part of the solution of the universe. If a universe making sense exists, then the tautological first principle says that there must be a space of this type. This is what we have been seeking. We have already explained why this universe must exist. Hence we can mark off one more self-authenticated feature. This universe must exist with a space.

Note that we started this analysis with the non-quantitative apeiron. Nevertheless there has to be a mathematical component of this fundamental reality. After all, mathematics is the descriptor of all forms of order. Furthermore we are programmed to seek explanations for order. Now, Fermat's principle exposes both the macroscopic order and the form of the mathematics. We should not be surprised that this mathematics is rather simple – the fundamental tautologies make it so. Indeed this fundamental simplicity is at the heart of the reason why everything about the universe is understandable.

This foundation space is the origin of the basic set of spatial variables known as coordinates. Hence this space is known as coordinate space. Of course this foundation space is not the same space that we observe everyday. The properties of these two spaces are quite different and anyway there is no matter in the foundation space. Our foundation space is only filled with aether. Thus we must take more steps before we sight that space which we think of as natural. However, our current aim was to uncover any space. Recall that we need this space to expose the source of universal logic. When we have uncovered this logic, we will have completed the *by inspection* proof.

Space Dimensions

Perhaps this result is not what you expected. Do not be put off; this space is more rational than our empirical space. What are we to make of eight spatial dimensions? Did not Einstein only add one time dimension to make four dimensions? Well let us think about the observed space-time in which we find matter. Is this familiar space really four-dimensional?

They say that fact is stranger than fiction. Did you know that in our world all matter is in continual motion? In fact everything is moving at the same speed, light speed. We do have a type of perpetual motion. No! I am not joking – this is conventional science. How is this possible? Well in relativity we describe 4-space as having coordinates denoting time and 3-place. (Note that to make intensions clear, we will prefix some words with their dimension.)

Suppose a bus is stationary in 3-space, then it is still moving forwards in time. Suppose the bus accelerates to a high 3-speed. The bus is now moving faster in the 3-place dimensions but it moves much slower in the time dimension. Indeed if the bus were to travel at light speed, then time in the bus would become frozen. For instance if a line of pedestrians could look through a window from the outside, they would agree that the passengers do not age. Actually a very strange sight would greet them. The passengers are going to look like cardboard cutouts! These are some of those intriguing novelties of relativity. Anyway this lesson also tells us that it is not enough to specify just 3-place and time. What happens in the material cosmos certainly depends upon the where and when, but it also depends on the motions of matter. Thus there are more degrees of freedom than those allocated by four dimensions.

Now, within the context of classical mechanics, Laplace pondered upon his own gedanken experiment. Laplace supposed that some great being knew the positions, velocities and masses of all matter at one instance in time and knew how they interacted. Then, by using the laws of physics, this being could predict everything that was yet to happen in the future. It does not matter whether this being actually exists, a being could hypothetically know the future. However, if someone could predict our actions, then there can be no free will. Well, nothing is freakier than an academic's wacky conclusion – especially one that no one can disprove. Anyway we will circumvent worrying about Laplace's conclusion until later. Here we are more interested in his argument's necessary inputs. Again we see that we need more free variables than just time and 3-place. We need to know about the mass and motion of matter.

Clearly these extra needed variables are 3-momentum and 1-energy (ie mass). Indeed we find pairings of 3-position and 3-momentum, and time and energy in many areas of physics. Thus we conclude that contingent arrangement of matter is also defined by eight variables. Thus our familiar space also has an eight-dimensional form. Of course the space of matter and the foundation space are otherwise very different.

Ultimately we can formally verify this coordinate assignment for the foundation space. We accomplish this task by viewing the form of the equations of quantum mechanics. However, let me show you why it is very reasonable. We must be able to factorize a coordinate infinitesimal. This localized double-algebra factorization looks like

$$\Delta a(\hat{r} + \hat{\imath} b)*(\hat{\gamma}_0 + v_1\hat{\gamma}_1 + v_2\hat{\gamma}_2 + v_3\hat{\gamma}_3)\,.$$

This formula contains the five real-number factors: $\Delta a, b, v_1, v_2, v_3$. The complex and quaternion directions are indicated by the carat symbol, ^. Let us not worry, at this stage, about units – everything is dimensionless. Thus we can directly compare the factorization result with the eight-dimensional coordinate representation

$$\hat{r}(\Delta\varepsilon\hat{\gamma}_0 + \Delta p_1\hat{\gamma}_1 + \Delta p_2\hat{\gamma}_2 + \Delta p_3\hat{\gamma}_3) + \hat{\imath}(\Delta t\hat{\gamma}_0 + \Delta x_1\hat{\gamma}_1 + \Delta x_2\hat{\gamma}_2 + \Delta x_3\hat{\gamma}_3)\,.$$

Symbolically ε is energy, p_i is 3-vector momentum, t is time and x_i is 3-vector position. Note that here we are treating these variables as independent. Thus we get what could be called the canonical relation

$$\frac{\Delta x_1}{\Delta t} = \frac{\Delta p_1}{\Delta \varepsilon} = v_1\,.$$

This result is at the heart of all physics. This is a spatial result. (Other familiar results apply to matter and not to space.) Furthermore it is clearly consistent with the standard definition of momentum as $p_1 = \varepsilon v_1$. Note that energy and mass are identical in our unit-less representation.

5. The Rational

We have built a vivid picture of the first elevated storey of reality. We have revealed a quantitative space. With its branch structure, this space is more general than a simple coordinate structure. We can fully describe this foundation space by using the language of mathematics. However, there is something else – the aether. The aether is an unavoidable companion for space. However, the aether is unlike space. Furthermore the aether is not in space (it is in the apeiron); yet the aether is irrevocably tethered to space. Thus space and the aether make up the first duopoly of the universe.

The aether has symmetries, which it acquires from being linked to and exclusively referenced from space. Like space, the aether has its own further levels ie its own higher aether with its own higher aether, etc. Thus the aether is an ordered hierarchy. Clearly the aether is non-numeric since if it were numeric, it would represent extra dimensions of space. However, the aether represents everything having any connection with space. Let me caution you that this foundation aether is not yet matter, energy, spirit, mind or anything that we directly experience. Nevertheless it is related to such concepts.

Logic

Clearly the scientific language of mathematics cannot properly describe the aether. We need some other representative scheme. We require something that is adept at describing the qualitative. As fully-fledged citizens of the universe, we already know of such a tool. We call this tool, natural language.

We shall explore this revelation further. It holds much to interest us. Now, languages are expressible in many different formats (eg speech, writing, codes, etc.). The details, forms and options of the aether are analogous to the script of a language. Hence we shall maintain that these details are the expression of a universal language – the langue. Now, please do not confuse language expressions and conversation. What we are saying is that this foundation aether attached to the foundation space is an unremitting hotchpotch of unintelligible grunts, groans and vacillations as interpreted in the langue.

Alternatively you may think of the aethereal field as a canvas upon which conversation or considerations could be imprinted by an intelligent system. Yet all that one can normally observe on this storey (of reality) is a discordant rumble. The foundation aether has only some of the potential for supporting practical thought. Primarily this aether lacks an accessible memory, and hence the aether cannot show learning, relational understanding, rational decisions or any consistency of opinion. Thus we will not discover the capacity of mind until we reveal the nature of matter.

However, I do not wish to tackle the fascinating topic of mentality yet. Those discussions are saved for later chapters. Our immediate aim is to complete our authentication of the apeironic form of the universe. Thus our goal is to show where the elements of the natural logic (available to all sentients) lie. To help with the explanation, we shall put some flesh upon the conceptual structure of the aether.

At this stage, venturing into too much detail about the writings embedded in the aether would be diverting. We shall discuss more details of the aether in the next two chapters. Therefore here we will simply illustrate the form of the aether by drawing direct parallels with the 4×2 dimensional form of foundation space. The fourfold symmetry delivers the basic structure of a reduced 4-space (space-time). Likewise the fourfold structure of the langue delivers the basic morphemes of the langue's vocabulary. The twofold spatial symmetry differentiates a coordinate from its sense of change. For example, in space, the twofold symmetry distinguishes momentum from position. As we saw, this identification is itself a natural consequence of the need to be able to factor locally the two spatial symmetries (complex and quaternion). Thus likewise we expect the twofold aethereal symmetry to differentiate a conceptual entity from its sense of action. Hence linguistically this twofold symmetry grammatically distinguishes verb from noun.

Now, both of the fourfold and twofold symmetries carry a directional sign (±). The symmetries must factor so that they become separable. For this reason the two symmetries must carry distinct directional signs. Thus the sign set that we use for the aethereal complex symmetry is not the same as the sign set we use for the aethereal quaternion symmetry. Note that the sign pair (plus and minus for a symmetry) obey very clear rules: (+) × (+) = (+), (+) × (-) = (-), etc. These rules are none other than the conventional relations of true (+) and false (-). This is just another way of saying that (-) signifies a reversal of sense and (+) signifies no change.

We will refer to a single aethereal expression as a semanteme. Thus what we have found is that we must have more than one sign accompanying each semanteme. Consequently a semanteme of the langue has one sign indicating whether the verb is true or false, and it has another sign indicating whether the noun is true or false. Hence we conclude that semantemes of the langue have vocabulary derived from the quaternion structure, grammar derived from the complex structure and logic derived from the signs.

Let me rephrase these conclusions. The aether is really just a random field expressing (in the universal langue) sentences built from elementary units of meaning ie semantemes. The semanteme is formed from elements of the complex and quaternion groups. Thus we identify this elementary structure with vocabulary, grammar and logical variables.

Hence we have uncovered the native logic of sentience. In this natural logic of the universe, a statement is not simply true or false. Essentially we have created four possibilities. A statement may be provably true, provably untrue, unprovably true or unprovably untrue. These extra options are simply a formal way of recognizing that some statements are disallowed or undecided in terms of classical logic. The words prove and unprove are used unconventionally. The standard meaning of prove is translated as prove-true in our new logic. The standard meaning of disprove becomes prove-false. Thus to avoid some verbal confusion, I negate the verb ‘prove’ as ‘unprove’ rather than as ‘disprove’.

The classic example of an unacceptable statement is *this statement is false!* This statement is a rephrasing of the liar paradox. In our enhanced logic, this statement is acceptably classed as unprovably true. Hence the statement is not seen as paradoxical. Look again at the statement. If the statement were true, then we have contradicted the statement’s claim. Hence we have unproved the assertion of truth. However, we can make no other claim. In contrast the futile statement, *this statement is true*, is actually unprovably false. It is impossible to demonstrate this statement false. We may only unprove that it is false.

Note that our new logic differs from classical logic. Classical logic has an axiom that states that every statement must be either true or false (just two choices). When we find statements that clearly contradict this axiom (as in the liar paradox) we are forced to exclude these statements from the scope of our logic. This is just like some early mathematicians who preferred to deny the validity of irrational numbers or imaginary numbers. Exclusions come at a great cost: they impede further developments; they create omissions and paradoxes. In contrast we maintain (from *ab initio* arguments) that every statement takes one of four logical values. Thus we employ no arbitrary exclusions. We allow everything to be accepted. This is even though some acceptance is seemingly senseless!

How did this new logic arise? Let us summarize the steps. Numbers and mathematics self-assemble in a few steps. We start with the presentation of integers (as sets of events) in the apeiron. Fermat's principle lets us construct a first-elevated storey of the edifice of reality. This storey has a foundation space built from real numbers. This space contains the aether, and subjects the aether to group symmetries. Through these symmetries the aether becomes a storehouse of linguistic opportunity. Furthermore the displayed symmetries harbour the elements and rules of logic at their core.

Thus in the scheme of reality, a clear logical mode that might be accessible by a sentient being, emerges before matter even makes an appearance. The rules and relations of logic are just the rules and relations of two pairs of mathematical signs assigned to coherent symmetries. One set of signs is associated with verbal grammar and the other set is associated with nounal grammar. These rules are a more general form of traditional classical logic. Nevertheless there is a clear natural logic and there is a justification for applying this particular logic.

Furthermore there are two aspects to our third weak assumption of an understandable universe. The universe must display logic and it must show features that make sense. This second aspect is what the first principle achieves. The tautological first principle endows the multistorey structure of reality with consistency. This first principle implants order and consistency ie something understandable in the world. Thus within this universe, each enfolding storey of reality (as built by the first principle) makes rational sense. Hence these features enable us to claim that we have rationalized the assumption of a sensible and logical universe.

Well, have you grasped the key insights? Perhaps you thought you had understood what I was arguing, but now I have gone and changed the rules of logic! Is your head spinning? Surely logic cannot be arbitrary. Can there really be alternative forms? Well, yes there can. However, there is one form appropriate for use by sentients. Furthermore we have just shown this most appropriate form is not that of classical logic. If your head aches, it is because conventional wisdom oozes with the fixed yes/no choices of classical logic.

Thus let us take a short break before concluding the derivation of a solution to reality. Let me take you to the fault zone of classical logic. Here I will apprize you of the failings of classical logic. Then perhaps this will make it easier to divest yourselves of those antiquated classical trappings.

Thus we shall start with the formal basics. We found classical logic on its three axioms:

- True and false are mutually exclusive attributes.
- If A and B are identical expressions,
  then A and B are interchangeable.
- Every statement is either true or false.

Unfortunately the use of classical logic produces a range of related paradoxes. As we have mentioned, 'This statement is wrong' is the simplest paradoxical example. If the statement is correct, then it states that it is wrong ie incorrect. Thus we have a contradiction. Alternatively if the statement is false, then the statement implies it cannot be wrong and so the statement must be true. Again this is inconsistent. Hence if the statement must be either true or false, we have an undeniable logical paradox.

To overcome such logical recursive problems, Bertram Russell (1872-1970) and others proposed using the 'meta' prefix. The idea is that we call a statement containing a reference to any statement, a meta-statement. Furthermore, we label an expression containing a reference to a meta-statement, a meta-meta-statement. This hierarchical categorization forcibly ridicules recursive loops such as: This statement is ineffable. The answer is we cannot know the answer.

Thus this scheme works by always quarantining the stated recursive loops in the wastepaper basket holding nonsense. For instance 'this statement is ineffable' is nonsense because it is a meta-statement and therefore the meta-statement has no referent.

Clearly the Russellian solution has some merit. It may help to visualize a concrete example. If you will give me a little licence, then I shall describe a tangible but fanciful physical example. Using a newly invented time machine, I have travelled back in time from the year 2124. Disastrously the machine inadvertently landed on my grandfather's pram and killed him. I am devastated; I will never time travel again. Regrettably neither my father nor I will be born. However, I am bewildered. Who killed granddad? Well Russellian logic would say that I am a referent to a history. Thus the act of travelling back in time creates a reference to an alternative history. Hence I must enter a different universe, a para-verse (parallel/paradoxical universe) belonging to a meta-verse system. I do not travel back to our universe at an earlier time, but to another previously equivalent para-verse. The version of me, born in our universe, killed the baby who was my grandfather equivalent in the para-verse. I am sorry, granddad. Russell says that logic cannot save you. Thus the paradox has been assigned to the bin. As a keepsake, I am left with a book that granddad gave me in the time-travel departure lounge, explaining the secrets of ultimate reality. Why did he do that? Anyway the scheme of 'meta' prefixes makes reasonable sense and even says that I can never go home to my version of the universe.

Thus this viewpoint is rational and is even suggestive of the multiverse theory. However, quarantine procedures (based upon denial) are never foolproof. Kurt Gödel (1906-1978) showed how one could sneak through the quarantine lines with a mathematical recursive loop. Gödel used numbers symbolically to represent mathematical expressions. Surely I should say meta-expressions? Well I could; indeed we need to say meta-expressions if we are to preserve the quarantine restrictions. Unfortunately mathematics is inherently a symbolic referential system. We construct mathematics to refer to itself rather than the outside world. We only *apply* mathematics to the real world. Indeed mathematics would fall apart and would be useless if it were cut into little pieces that could not work together. For instance, look at the way I defined a number and its algebraic rules in a circular fashion. Look at the way we define integers by recursively adding one. Look at the way we analyse and classify numbers as irrational, integer or prime. Look at the way we take a real part of a complex number. We cannot afford to add a 'meta' prefix for each and every high-level referential operation. Hence there are no acceptable meta-aspects in mathematics.

Thus Gödel had indeed opened *Pandora's Box*. He was able to analyse mathematical proofs derived from axioms, in terms of the mathematics of the axioms. Then, he introduced a paradoxical formula that effectively asserted: the formula corresponding to number, n, cannot be proven from the axioms of mathematics. Of course the number, n, is chosen to correspond to this same formula. Hence he brought into the open an unprotected self-reference. If the expression is false, then the expression can be proven from the axioms of mathematics ie the expression is true. Thus we would have a contradiction. Alternatively if the equation is true, then we cannot prove it. This conflict is unpalatable but inescapable. Consequently classical logic requires that the axiomatic basis of mathematics is and must for ever remain incomplete because it cannot prove something that is true. This enlightening result is known as Gödel's theorem.

Of course there are other possibilities: mathematics could be wrong (it is kinder to say the axioms are inconsistent) or classical logic could be wrong (it is kinder to say the axioms are inappropriate). Let us consider whether classical logic could fail. Luitzen Brouwer (1881-1966) suggested that we should re-examine the rules of logic. We should not accept the word of Aristotle; classical logic is not self-evident. Thus Brouwer founded an approach to mathematics (intuitionism) denying that every statement must be either true or false. Hence, in his formalism, he kept the first two logical axioms and dropped the third.

Interestingly, the sentient logic that we have uncovered is precisely of this form. However, we have specifically shown that there are always four logical values. Each logical value corresponds to a distinctive result. The table gives examples of statements belonging to each of these categories. We have also shown how our new logical categories relate to the classical categories.

Table 3: Examples of self-referential statements belonging to each of the four distinct logical categories.

	True	False
Provable	This is a statement.	This is not a statement.
Unprovable	This statement is not provable true.	This statement is provable true.

Hence, in our natural logic, we do not have to prove every statement is either true or false. Thus we are happy to accept the Gödelian self-referential expression as unproven true. This approach does not mean that we take the Gödel expression to be true (ie proven-true). As in a Scottish court of law, unproven does not assign a definite verdict. Hence mathematics is *not* shown to be incomplete, just *limited* in scope. Note that primarily mathematics attempts to deduce two things. Are provable expressions true or false? Are expressions provable or unprovable? These unprovable options are the new form of the Russellian quarantine zone. We do have somewhere to put the troublesome referential statements, without an unwarranted 'meta' layering of logic and mathematics.

What does this mean for my granddad? Do we still quarantine his premature death to lie within another universe? Well the new quarantine rules are not as strict as those laid out by Russell are. We do not have to quarantine all high-level referential processes. Consider the sentences: This English sentence contains six words. This English sentence contains forty-two letters. Obviously we should accept both these sentences as proven true. (Russell's classification would label them nonsense emulating a valid result.) As an aside, let me note these self-consistent statements are true for the same reason the *by inspection* proof is true.

Thus high-level referential statements can be true. Nevertheless we must quarantine the paradoxical and the senseless references. Thus if time-travel into the past is permissible, we would classify the new paradoxical universe as unproven-true. Hence if we send some voyagers back in time we can never know what happens to them. Do they make it into the past unhurt? Does the unresolvable conflict of logic obliterate them? Well, the unproven status of backwards time-travel must remain for all concerned: for us, for the time-travellers and for the inhabitants of the alternative universe. We do not even have to invoke Ockham's razor to see the difficulty. Their logical existence cannot be preserved in another universe. Thus on the worldview level of classical mechanics, logic forbids limited-impact time-travel into the past. Stop and think for a minute. This is one way of explaining why we, as macroscopic beings with a logical awareness, have a one-way time sense. Classical causal logic has a strict time-sense; classical deviations from a fixed time-sense, fall into the quarantined bucket for the certainly indefinite.

I have a good reason for putting in a few qualifiers here. Perhaps it is possible to rewrite the entire history of the universe whenever a time loop is attempted. Image, if you will, that each time I attempt to travel back to granddad's childhood, the universe restarts with granddad in his pram. Quantum uncertainty ensures that history will not repeat exactly as before. Maybe, next time, I will pay attention to that book that granddad gave me. Thus I decide that time travel is too risky. Hence I stay put. Then the universe can proceed without having to regurgitate the same loop (*a la* movie 'Groundhog Day') another time. This is an excellent outcome: I survive and Granddad lives.

Furthermore the logical conflict only arises on those higher storeys of reality, which guarantee the certainty of classical phenomena. Although I am not going to discuss it in this book, time-travel into the past (on some branches of space) is a necessary and natural part of quantum mechanics. Indeed the 'Groundhog Day' analogy is apt in the quantum realm.

QED!

Well there we have it. The revelation of the sources of rationality completes the *by inspection* proof. Thus we have presented the rationale for everything that is. A rational reality bootstraps its way into existence. The universe is not only a logical necessity; a logical tautological loop gives expression to its fundamental description. Thus one might even say that we have given a *reductio ad tautologiam* explanation for the universe. This derivation justifies everything about the universe.

Note that by completing this *by inspection* argument, everything and every step of the argument is self-authenticated. The universe must exist because there is an unlimited supply of monadic events. Nothing prevents any event from pointing: we have no constraining inputs. Thus we cannot stop the abundant monads from pointing. Hence the unlimited apeironic network must exist. As a (real-estate) salesman would say, the apeiron is an opportunist's delight. It provides the firm ground upon which any renovator can build a reality. Indeed intelligent fortune smiles on the higher storeys of the edifice of reality. Thus space, logic and reasonableness (as it applies to laws not people) are simple properties of these higher storeys.

Amazingly the arguments justifying reality do not rely upon external inputs like logic, mathematics or even the existence of the universe. The self-consistency criterion ensures that the developed solution of the expanded problem (with weak assumptions) is a suitable solution to the no-input problem. However, this solution demands that the universe of this solution exists. Consequently the argument is shown to validate the existence of the universe. Furthermore, this universe cannot take any form – it must be a special universe built upon the apeironic network by the first principle.

So what have we actually decided? Well the fundamental universe is an indefinite assemblage of references. The references (events) just say "that". Furthermore each "that" only refers to another reference. Hence no single "that" means very much. It is the pattern of "thats" that matter. This pattern lets us built up structures of information and meaning. However, the endless possibilities are generally self-defeating. Thus we start to look for order in the ceaseless din. We find this consistency by scaling up and looking for new forms of cooperative stability. This is Fermat's principle in action. We seek those features, which will be dominant. These new features are not simple references anymore. These features are higher-storey constructs. Hence, on the first-elevated storey, we find a duopoly of space and aether. On this scale, space and the aether are everything; there can be nothing else. Symmetries provide the certainties and the rules; the rest of the details are contingent.

Now, this construction is only the beginning. We have simply reached the first-elevated storey of reality. However, it transpires that each time we climb an extra storey, the viewing scale becomes larger and the view becomes more certain and more predictable. In this work we shall not climb very high. Let me remark that I have shown elsewhere that on the second and third storeys we find the two parts (state evolution and measurement) of quantum mechanics. On the fourth elevated storey we find the beginnings of classical mechanics and full determinism. This is how physics tallies with the multistorey structure of the edifice of reality.

The operation of the first principle is of prime importance. Thus let me explain again just how it is possible for the opportunist apeiron to be transformed into the definite world of experience. We shall consider a gedanken experiment involving the tossing of a coin. The outcome of the next toss is undecidable. We will not impose any bias on the coin and hence a result showing either a head or a tail will be equally probable. The rules of the coin toss do not change if we put the head imprint on the tail side and the tail imprint on the head side. This is an example of symmetry between the heads and tails outcomes. Let us toss the coin a few times. The outcome is unpredictable. Indeed in any small sample of tosses there can be big variations in the fraction of tosses displaying head-side up. Yet as the number of tosses increases, the fractional variation in the result decreases – the average result becomes more and more certain. This is how order and certainty magically seem to appear.

Perhaps you recall my earlier remarks that whether something and nothing are mutually exclusive, depends upon your viewpoint. One must always treat negative definitions carefully. A similar circumstance applies to the apparent exclusivenesses of order and chaos. When we use the same descriptive parameters, an ordered system is clearly not chaotic; likewise a chaotic system is not ordered. However, when we apply the first principle we look for a new stable set of descriptive parameters. This new description exhibits regularities appropriate to this new certainty. From this new viewpoint there is order in chaos. It may seem hard to accept at first, but every statistician knows that we depict disorder in an ordered way.

Now, there are several important points that this coin-toss example illustrates. We have to change the scale. We do not look at one toss but thousands of tosses. Furthermore, we change our basic constructs. We look for what is stable in the macroscopic system. The fraction of results that show heads is stable. Thus we stop talking in terms of the outcomes of single tosses. Instead we talk about the average outcome of very many tosses. Inevitably every time we use the first principle in this way, we discover a whole new more ordered and seemingly more coherent description of reality. In principle latent macroscopic order is hidden everywhere. All you need is a good tautology (the first principle) to unbury it.

Finally let us reassess and summarize the *by inspection* methodology. The statement, that there can be nothing other than the universe, is the definition of the universe. The statement does not imply that anything exists. However, whatever does exist is called the universe. We re-express this definition in a different way. We state there is nothing else that can have any influence upon the universe. Then, we add a tautological loop to our definition. This loop re-interprets our definition and delivers the rational picture of the apeiron. However, all these arguments apply in reverse. We can take the apeiron and deduce that the universe is everything. Thus we observe that the apeironic model is nothing other than a more elaborate way of expressing the original definition of the universe as everything.

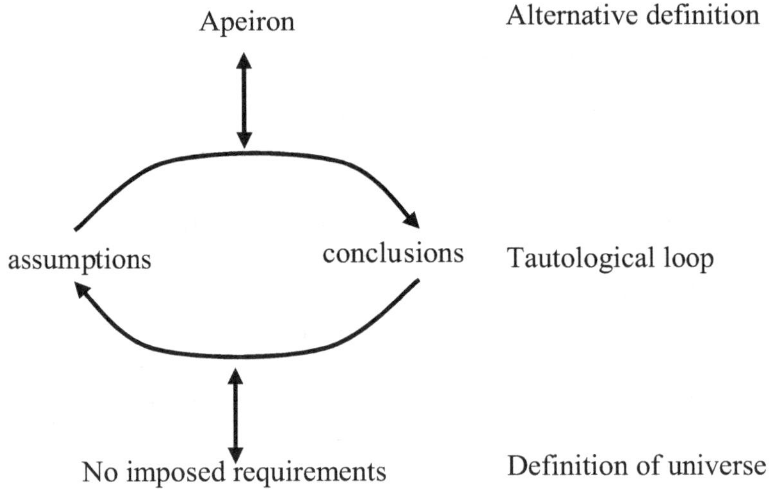

Figure 3: The universe and consequently rationality and space are just aspects of a fundamental definition. We can describe the by inspection proof as simple definition + profound tautology ≡ profound definition.

Banishing Scepticism

In recent millennia, philosophical inquiry has been weighed down by an inadequacy. The universe could not be rationally justified. Further, there was apparently no significant hope of a justification. Hence it seemed we could not know what natural form the fundamental universe takes. Therefore some thinkers have got away with proposing quite outlandish ideas. Moreover, other philosophers just like using the undeniableness of such unlikely ideas to attack the usefulness of arguments proposed by their colleagues.

Therefore there have been many suggestions deprived of sensibleness. I shall list a few. There could potentially be nothing other than the real sensory data of one's own nervous system. Thus there would be no external physical world. Alternatively even my sensory data could be a hallucination. Then, only my immaterial mind would exist. I just imagined the sensory inputs. Hence there would be no physical world at all. However, even if a physical world exists we could not know that a trickster did not create it at 08:00 this morning – with a complete set of archaeological records, personal memories, artefacts, etc – so that we would believe that the universe had existed since an imagined big bang. Then, there are the more pernicious suggestions involving zombies and supernatural spirits.

Fortunately such unwelcome sceptical positions are now untenable. We know the fundamental nature of a self-justified universe. We do not need and we have no freedom to impose further restrictions upon the entire universe. Clearly the discussed sceptical positions involve making additional unjustifiable propositions. Thus such sceptical stances have become less tenable than that far-fetched view that sentient aliens have visited Earth. At least the alien hypothesis is conceivably possible within the current paradigm.

We have also banished another important source of scepticism – doubts about a font of reasoning. It was known, in ancient times, that two equally reasonable arguments could give contradictory conclusions. Philosophers asked: Could there be a beginning to time? What begot life? How can we justify reality? From what are things made? Is there an ultimate cause of everything? Do we have free will? Is there more to the universe than that we can sense? Why is the universe so understandable? Will we ever be able to explain everything? As you can imagine, when faced with such fascinating questions, the great thinkers of history came up with carefully reasoned answers. Unfortunately the answers were often at odds.

In 1781, Immanuel Kant (1724-1804) assembled and analysed several of these conflicting arguments. These pairs of contradictory arguments (thesis and antithesis) are called antinomies. Now, the underlying issue (behind these antinomies) is really how we justify anything. Suppose a rational argument provides a reason for something. This given reason is justified by some other reason and so on. Unfortunately if this chain of reasoning goes on for ever, then we never have a completed explanation for anything. Thus it appears that the chain of arguments has to stop somewhere. The only way anything can be justified is with a foundation to the chain of reasoning. This conclusion is the basis of the argument presented as the thesis.

However, the antithesis points out that a chained argument cannot stop arbitrarily. There needs to be a reason for why it stops. Thus we still have something to explain. Further it appears to be contradictory the starting reason has to justify itself. A self-justifying premise (tautology) would be an empty premise without useful content. An empty premise justifies nothing. Thus if the chain terminates, then the chain would seemingly be left unjustified. Hence there can be no reasonable end to the chain of arguments. The chain must go on for ever. This conclusion is the basis of the reasoning known as the antithesis.

Now, each of Kant's antinomies has its tailored variations. However, you will have got the basic idea: the chain of arguments is damned if it does have an end, and it is damned if it does not have an end.

So how does eventual reality overcome the dilemma? Well, we use a rational loop that lets us handle worries about logic, existence and descriptions together rather than separately. However, the reasoning does not start with this tautological loop. The argument starts with the no-input requirement. This statement is a definition of universe. This statement is self-supporting but it is not a hollow requirement. It says that there is no bias or favouritism. We cannot insist on any principle, law or observation about the universe. We cannot say that we must have a space of 1-time and 3-distance dimensions. We cannot say that there must be mentality or matter. We cannot dictate the values of fundamental constants. In fact, the list of things that we cannot impose is endless. Everything that we cannot impose becomes an important explicit condition. Hence we have an unstoppable stream of explicit information about the universe. This stream is contained within the single implicit no-input requirement. Thus we have a useable beginning for our justification of a chain of arguments. Furthermore this logical starting point translates into a well-defined foundation storey for the multistorey edifice of reality.

Finally let us mention science and science's practical scepticism. This scientific scepticism is of a different type. When scientists say that they are sceptical they mean that they want to test the conjecture. They want to see practical evidence and confirmation of proffered propositions. This is because science is cautious. Science has self-improvement mechanisms to enable it is grow. Science has self-repairing mechanisms to correct errors when it goes off the rails. These are the practical applications of scientific scepticism, the careful investigation of a provisional doubt. Thus scientific scepticism is not the dogmatic scepticism born of philosophical despair that I have been describing earlier. Such doctrinal scepticism dates back to the Greek philosopher, Pyrrho (365-275 BC). Whilst few, if any, modern philosophers are outright philosophical sceptics, they are still apt to use sceptical arguments to disparage the views of their rivals. Hence removing any support for such doctrinal scepticism is long overdue.

Obviously scientific scepticism of the hypothesis of eventual reality is healthy. The theory needs sceptics to test its predictions. As they say, exceptional claims require exceptional proofs. Unfortunately predicting the forms of already known physical science does not provide overwhelming evidence for the theory. Thus there is much further exploratory work to be done to achieve scientific approval. However, we should note the hypothesis' rational justification is without assumptions. Hence eventual reality is beyond doubt the archetype for all scientific theory. Thus eventual reality will win any beauty contest between hypothesises. Likewise Ockham's razor must select eventual reality as an explanation for reality. No other theory could be simpler. Furthermore eventual reality is the first theory to embrace both matter and mind. Thus I suspect the most direct and convincing experimental evidence for the theory's validity may well emerge from the still immature studies of mentality.

End of Reductionism

In science we often try to understand tangible things by finding out what is inside them. For instance, the human body contains many processing systems: respiratory, cardiovascular, digestive, lymphatic, excretive, reproductive, nervous, etc. These systems contain functioning organs. The organs are made from functioning cells, etc. With each step, the level of complexity decreases. We call this process reductionism.

Likewise in physics we have categorized the observed material world down to a bunch of elementary quarks, leptons and force carriers. (I am deliberately ignoring the Higgs boson and other speculative particles.) We often compare reductionism with a set of Russian dolls. We open one doll and there is another smaller doll within it. This exposure repeats until ultimately we find an innermost doll.

Reductionism is not to be confused with the regression of worldviews. When we discover new material particles within other material particles, we do not necessarily change our general beliefs about operating laws. We would have to describe the world differently to warrant a change in the worldview. Thus (if particles continue to obey quantum mechanics) we might need to discover something else (other than other particles) lurking within a particle before we need a worldview change. Only then will this regression necessarily accompany a reductionistic change.

Now, reductionism has been very successful. Nevertheless there have always been some philosophical doubts about whether it would be fruitful to continue to delve down ever deeper into the make-up of reality. Could a whole system possibly be greater than the organization of its parts? Does reductionism lead us to overlook some holistic issues? We can appreciate the supposed dilemma by considering any practical machine that we have designed and built. The machine operates according to the laws of physics. We can examine how the machine works. However, the laws of nature do not address the reason that the machine exists, the choice of materials used, the machine's relevance or its contingent aesthetics.

Thus we can describe a machine in terms of its parts. Yet this reductionist description misses any understanding of subjective and qualitative elements. Perhaps we are not too worried about a machine that we have built. However, we are concerned about deep philosophical issues: the reason the universe exists, the composition of the universe, the universe's relevance and the universe's appearance. Clearly the scientific laws of nature do not answer these questions.

Fortunately reductionism is only about the outwardly tangible details. Science and natural philosophy are more broadminded, they have a quadripartite form. Hence the explanations (including roles) of all things may be given by the belief structure of the worldview. The worldviews have their own regression.

Now, it goes without saying, we will not understand ultimate reality using a higher-storey picture. Inevitably the picture of the self-justifying foundation must be different from pictures for higher storeys of reality. These higher stories draw their characteristics from lower storeys. Thus we can argue that we need a final change in worldview to accompany the final reductionist simplification of basis form. Then, we hope to have an accompanying explanation for existence, composition, relevance and appearance.

Of course this is what we have already revealed. There is a rational foundation. We have been able to reduce reality to this foundation. Surprisingly the foundation describes the fundamental nature of the universe including its necessary property of existence. The foundation description of reality is unique; the foundation is entirely made from idealized forms that are not observed on any other storey of reality. This foundation has no input constraints.

Hence we have completed the major conclusion of this document – a rational explanation of everything. Now, in the majority of the rest of this book, we will discuss how science relates to the concepts of this new hypothesis.

6. The Body of Science

Start to Science

To suggest the reports of the end-of-science are premature is to acknowledge that there is an impending end. However, eventual reality shows only a start-to-science. Science remains open and boundless. Thus it is fascinating to see how this start-to-science overcomes the known apparent incongruity of a beginning to science.

Principles of Nature	**Fundamental Entities**
Initial, boundary or identification conditions	**Properties or characteristics of fundamental entities**

Figure 4: Quadripartite features behind searches for a unified foundation to science.

A satisfactory beginning seems impossible within the conventional quantitative mindset. It is not sufficient only to determine a single fundamental principle of nature from which all other principles and laws can be deduced. We need to specify a four-fold structure. This structure details the constituent entities of fundamental reality, the properties classifying these entities, the foundation principle and a selection (eg constants of nature) identifying which of all conceivable universes we experience in practice.

However, the resolution is simple. We must stop thinking about things and start thinking about logical concepts. Clearly the ultimate fundamentals do not belong to the scientific world. The quantitative will not explain the use of or form of logic. Nevertheless in the metaphysical realm all aspects can be tautologically reasonable. Hence we were able to uncover a quadripartite solution. The entities are featureless monads. The monads are associated by references. There are no (other) laws about this pointing. The network endlessly continues.

Then we took this foundation, effectively added the Planck scale and applied the first principle. Naturally the foundation of scientific reality is the quantitative. The fundamental quantities are the elements of symmetry groups. There are no (other) restrictions on the expression of these symmetries. We view this foundation system as a multi-branched space with aethereal field. The aethereal hierarchy continues indefinitely.

Necessarily this scientific foundation has conceptual inputs allowing its creation. Thus we build the scientific base upon a pre-existing metaphysical reality with a pre-justified logic. Hence we can construct an explicable foundation for science (on the first-elevated storey of reality). Critically we add the foundation unit of all measurement as a device that creates the quantitative. However, this unit has a clear rational representation. Interestingly in our higher-storey space, this foundation unit is still relevant in calibrating the dimensional size of the observable universe.

Matter

Now, at this stage in reading the book many readers will appreciate just how much we have achieved. However, some minds will be reeling. They will worry that we might not have made enough convincing progress. Certainly we have given a credible *ab initio* argument for the existence of a universe. However, a pessimistic sceptic is likely to say that they do not recognize this universe. Hence there must be something wrong with the argument. After all, the foundation space is boundless. The aether is qualitative. Neither member of the duopoly is at all familiar. Fortunately we are not lost in a fantasy of philosophizing. This unfamiliarity simply explains why the deep thinkers of the past did not discover the hypothesis of eventual reality. There is a giant gulf between the apeiron and the familiar classical world. However, we can breach the gulf.

Not surprisingly the analysis becomes quite involved when we come to the technical details of science such as: the constants, laws, theories and principles. Hence, in this work, we shall only give brief mentions of how eventual reality merges into the known science of matter and space. However, first let me try to put this dismissive urge in context with some examples.

Imagine that it is Vienna towards the end of the nineteenth century. The restaurants are alive each evening with vibrant intellectual discussion. The eminent Scots physicist, James Clerk Maxwell, had earlier derived the probability distribution of molecular velocities in dilute gases. Ludwig Boltzmann (1844-1906) was building on Maxwell's work. Boltzmann was working on how chaotic behaviour increases in systems of interacting particles and how energy was shared between the various avenues available. Boltzmann was engaged in pioneering theoretical work; he was making seminal strides at the forefront of theoretical physics. However, two entwined problems impinged upon his endeavours. He suffered severe bouts of depression, and stringent critics disputed his fundamental premise ie that atoms and molecules existed. If atoms and molecules did not exist, then his work was irrelevant nonsense.

This clash of values is part of a much older dispute, a dispute about the reality of intangible things. Your personality helps decide your natural bias in this debate. Many people find abstract concepts very difficult to grasp. Probably most people prefer to think about the tangible (anything that can be detected by the senses of a man or machine). Let me give an old example to illustrate the style of argument. Zero is the number that represents nothing! In the past, many have claimed that we do not need a zero: How can we need something to represent nothing? Thus these people did assert that zero does not exist. Perhaps you can now appreciate why zero was such a late addition to our number system. However, there is only one possible outcome to this reoccurring generic dispute about the intangible. Will you permit me another vacuous statement? It is not useful to deny the existence of something that is useful. Hence every useful concept is a sensible part of reality.

Thus the proponents of a reality rigidly clinging to the tangible can never win their argument. Yet their strident contribution to a debate will sometimes force a compromise position that all debaters reluctantly accept. In reality any compromise position is ultimately found to be untenable – one cannot be half pregnant. Perhaps the interpretation of quantum mechanics provides the clearest example of an uncomfortable extant compromise. Thus the quantum state holds all that we can possibly infer about (classical) reality. Yet the quantum state lacks the definiteness needed to describe every feature of (classical) reality. Hence by normal convention we only attach the word, reality, to measured or otherwise known things. Recall that I avoided this equivocation by introducing separate notions of reality (for different storeys of reality) and thus allowed a tangible classical reality and a less tangible quantum reality.

Clearly doubting Thomases are in good company but their disquiet is unwarranted. We can show that an abstract apeiron holds a sequence of realities that gradually build to one, which is fully concrete. Next, we will indicate how we fill in the detailed aspects of the gap between conventional physical science and the first-elevated storey of reality. Fortunately our thorough analysis has provided the tools needed to construct each storey of the edifice of reality. Hence on the second-elevated storey we do find the first appearance of quantum matter.

To proceed we again use Fermat's principle to build our picture of the universe. Familiarity with the science of the universe is a boon. Thus once more, we ask what macroscopic style of picture we want to construct from our starting (microscopic) picture of space and the aether. Hence let us start by asking how we might recognize the stuff called matter. What is it about matter that makes it matter?

Well, imagine slamming your hand into something solid. We are certain that matter is there. Matter is not something arbitrary. It is not a hypothetical concept. Close your eyes and open them again. Matter is still there. Matter is definite. Secondly we know that science successfully describes matter. Furthermore science deals with the quantitative. Thus matter is undoubtedly quantitative; otherwise we would have had to define science differently. Thirdly observed matter is made from more fundamental discrete units (essentially particles). However, this observation does not rule out string theories or advanced spatial theories; it is merely recognizing that matter comes in chunks.

However, we should not follow the lead of earlier philosophers by overstretching our ideas. Hence we will not imply that everything about quantized matter is precise, quantitative, definite and certain.

Rather, we are saying that the existence of matter is certain and the description of matter is quantitative. Nevertheless any stuff, which is both certain and quantitative, is the stuff called matter. Note that I am not distinguishing matter from other forms of energy (eg radiation). As required by special relativity, we shall regard matter and energy as being the same primary stuff.

Obviously the qualitative aether does not directly satisfy this requirement. However, the aether is all that exists in space. Thus we have to suspect that this field hides both matter and mind. Now, the aethereal field resembles a macroscopic junk heap of microscopic pointing references. For the most part, there is neither rhythm nor reason behind the pointing choices. Such pointless pointing creates bedlam (and a bad pun). Thus this chaos is another way of saying that there is no intelligence (consistent choices) displayed by most matter.

In principle, we can measure the extent of this disorder. This measure is given by a field called the entropy parameter. This quantitative parameter does not have to be the entropy as such, just something like the entropy. For instance the entropy parameter could be any monotonic function of the entropy eg the exponential of the entropy. The entropy parameter is the only definite statistical measure of the chaotic aether. This entropic field is unlike the aether. The entropy parameter describes a quantitative field that has clearly become an abstract property of space (in the same way as a shadow becomes a painted feature of the surface upon which it falls). Furthermore this entropic field has shrugged off the aethereal pointers. However, it takes on a multi-dimensional form as conditional cross-correlations of orderliness across spatial branches.

Hence to reveal quantitative matter, we start with the quantitative measure of the aether ie the entropic field. Then, we use the first principle of the universe. However, as befitting a tautology, the first principle is only guidance. We must seek the coherent constituents of the entropic field. These constituents will be stationary forms. Now, there is no reason to suppose that the microscopic entropic field is analytic in space. Yet when we rescale we loose all microscopic detail and the macroscopic form will become analytic. Then, we see coherent order in the macroscopic stationariness.

To picture what we are doing, consider an expanse of air. The air's molecules are busy heading in all sorts of directions at all sorts of speeds. Microscopically we cannot recognize any sense in their behaviours. However, on rescaling the picture and seeking the coherent attributes we can discern the melodic sounds of music. Mathematically these sounds are picked out by performing what we call a collective analysis – an analysis that looks for an accumulation of cooperation. The result of this analysis is a set of amplitudes, frequencies and phases of the pure notes. Thus this collective representation describes the macroscopic coherent form as a combination of discrete units (musical tones). We are seeking just such a format to reveal a distributed system of discrete matter. This is why we perform a collective analysis of the entropy parameter.

At first glance, there appears to be a serious technical problem. The entropic field with its correlations across the branches obeys complex-quaternion algebra. This is the same algebra as satisfied by the foundation space. Yet our favoured collective analysis only generally works for commuting numbers (like real numbers). Nevertheless the problem will turn into a blessing in disguise – the solution will engineer a generous bounty of scientific capabilities.

Clearly this algebraic issue is quite important. So let me make a series of digressions to explain how we solve such problems. Firstly let us talk about real and complex numbers. We can assemble a set of algebraic axioms obeyed by real numbers (but not complex numbers). However, as we noted, it is useful to have a number system that goes beyond the real numbers. Hence we also added the imaginary numbers to deliver the system of complex numbers. Complex numbers have a slightly more relaxed set of axioms than real numbers. (Real numbers can be ordered according to size.) However, imaginary numbers do not form an independent number system. We add imaginary numbers as an extension of the real-number system. This illustrates how we may defeat an annoying axiom. Simply we append an extension to our number system. Then, we can both retain the old numbers and remove unwanted complications of an axiom.

Thus we take the complex-quaternion algebra and append a dependent system of quasi-numbers. These numbers are given properties that will enable us to perform a collective-coordinate analysis. However, it is not immediately clear what axiom we should relax. Certainly we cannot reimpose commutation. To appreciate the actual requirement I want you to remember how slide rules and logarithm tables work. They work because exponentiation relates addition and multiplication. Hence we can replicate multiplication by performing an addition and transforming the numbers. A table of natural logarithms (or the markings on the slide rule) performs the necessary transformations. Formally we write this relation as:

$$\exp(x+y)=\exp(x)\times\exp(y)\,.$$

Now we want this same mathematical property in our analysis but we only need the result to apply to the product of quasi-numbers and complex quaternions. We find that we can achieve this requirement by relaxing the axiom that demands numbers have inverses.

The outcome is quite surprising; we get a double extension of the algebra. Quasi-numbers must obey the algebraic rules. Furthermore we only want the product of quasi-numbers and complex-quaternions to satisfy the exponential requirement shown in the equation. Thus we make this product part of a further extension of the number system. Such extra numbers that anticommute and satisfy the exponential requirement, inevitably are Grassmann numbers. However, we will not be pursuing this mathematical analysis of matter much further here. Interested parties can find the mathematical details elsewhere. What we find is that the product of quasi-numbers and complex-quaternions form their own closed axiomatic set of algebraic rules. These rules are the axioms of a Grassmann algebra. Physicists have already established that Grassmann algebra is the algebra of fundamental particles. Hence that bothersome algebraic problem turned out to be the source of fermionic statistics.

Next let us turn out attention in a new direction. In preparation for an explanation of mind, we shall look at what are the basic features that we need to explain.

Empirical Mind

We all have minds and hence we appreciate a great deal about mentality. Let me run through a few issues. The Greek anatomist, Herophilus (335-280 BC), first appreciated that the mind and brain are linked. However, not that long before this, Aristotle had taught that the mind lay in the heart. Anyway today there are many indicators that the body and in particular, the brain influences the mind. Chemicals circulating in the blood will affect mental mood. Damage to areas of the brain can destroy or impair some mental capabilities. Fortunately brain plasticity can lead to a limited restoration of some damaged mental functioning. Physical sickness can modify people's personalities. Observable brain aging alters mental capabilities.

There are also clear indicators that the mind influences the body. We have free will and can choose to move parts of our body. A placebo can quicken patient recovery. The mind is open to hypnotic suggestions. Witchcraft, voodoo, pointing of the bone and other superstitious beliefs can sometimes produce physical outcomes.

So what are mind and brain? Is mind a property of the brain? Are the brain and mind distinct objects?

I will not bother saying much about that lump of grey matter known as a brain. Neuroscientists are still learning about brain organization and activity. Let us concentrate upon the nature of the more illusory mind. The mind is the seat of consciousness, yet the mind is more than the conscious state. We do not have to be in a self-aware state to have a mind. Clearly we still have a mind even when we are asleep, anesthetized, sedated, hypnotized or comatose. In these states, our stream of conscious awareness is interrupted. Nevertheless mental processes continue.

Furthermore even when we are conscious, there are many functions being processed in the background by a subconscious. This subconscious can respond instinctively to catch a ball, jerk a knee and initiate laughter. We recognize this subconscious as that part of the mind that does not possess an untamed will. The response of the subconscious is mostly a combination of learnt (ie programmed by the conscious) and innate actions.

The brain is part of the nervous system. Damage to the brain can lead to mental impairment. Damage to other parts of the nervous system does not appear to affect significantly mental functioning. However, there is one contrary affect – an amputee may experience a phantom limb. Nevertheless collectively this evidence supports an association of the mind solely with the brain.

Now, each part of the nervous system performs specialized functions. Its ability to save and recall memories uniquely specifies the brain. It is likely that this storage feature is the essential characteristic that differentiates a healthy functioning brain from other parts of the nervous system. The subconscious makes use of a short-lasting mode of memory. This memory allows us to dream and then possibly remember the dream if we return to consciousness soon enough afterwards. On awaking, details of any dreams recalled re-enter the stream of awareness and we may reprocess this data.

We have already noted that minds demand a memory. A mind could not recognize anything without a memory. Furthermore a mind could not hold a consistent view on anything without a memory. Fortunately active minds are embodied with memory and hence the mind can strive to achieve any purpose. Otherwise mindless (memory-less) matter is gormless, and the consequences of statistical chance will flourish. Interestingly Greek mythology mirrors this core role for memory. Mnemosyne, born from the union of earth and sky, was the Goddess of memory. She was the mother of nine Muses who inspired creativity across the arts and sciences.

The subconscious is machine-like, processing and regulating programmed bodily changes. This role of the subconscious is vital, allowing the conscious to concentrate on higher-level thoughts. Yet the distinction between conscious and subconscious activities can become blurred. At will, some people can consciously control a (normally subconscious) bodily function such as pulse rate. Furthermore an experienced driver may carelessly hand over a complicated conscious function like driving a car, to one's subconscious.

Thus we conclude that minds are always intimately linked to parts of matter called brains. Brains are major organs in the nervous system of higher living organisms. A brain is that part of a nervous system that possesses an internal memory. The brain maintains memories and can read and write memories. Machine brains exist and they can simulate intelligent choices. However, we attribute minds only to those brains that show genuinely free-choice behaviour. Minds have two functional components: an automated subconscious component and a free-choosing component that can become conscious. The subconscious mind is simply the functional part of the brain. It executes quite complex programs. Some programs run autonomously, other programs run on request from the conscious.

A conscious mind has a temporally continuous stream of holistic experiences that are fed by the brain's codified interpretation of input sensors and other mental functioning. This conscious awareness is not a collection of discrete experiences; it is one integrated experience at each moment of awareness. Furthermore this holistic experience is more than information; it contains complex associations including feelings, comprehensions, recognitions and active intentions. To avoid clumsy talk of a conscious in an unconscious state, I will refer to the part of the mind that can be in a conscious state as the higher mind.

The unconscious state of the higher mind performs general housekeeping tasks but also idles through a list of uncompleted tasks that need greater resource allocations than are available during conscious operations. Presumably this is why sleeping on a difficult problem can often help.

Thus conscious experience is of the form of a continuous stream of involved awareness. This stream is subconsciously analysed with only these analytical impressions sent for storage into memory. Without an eidetic memory, we cannot recover much of an earlier experience. Choices, feelings and meanings can accompany conscious experience. This stream of consciousness is a blend of subconscious inputs that are initiated by the senses and by automated or programmed routines.

Now, as we look further into details we find more clues to the mind. The contents of our conscious awareness are only a mental construction. Let me give an example. Suppose we look at a shiny red ball. In our appreciation the ball has a vivid red colour. However, this colour is a simple codified description of the detailed spectrum of light emanating from the ball in the physical world. Most people have four types of light sensors in their eyes (rods and three types of cones). Thus the colour codification of the world is our construct of an analysis involving four sets of signals. Other animals have quite different light sensors; hence they codify the appearance of the world quite differently. For instance, most butterflies have a colour vision that ranges into the ultraviolet and most butterflies can probably even distinguish between light polarization directions.

However, this distinction between a codified conscious awareness and the topic of its focus, applies to more than vision. Every aspect of our mental awareness is a codified construct. Even our awareness of our own thoughts is achieved by a codification, known as human language. Generally we are not consciously aware of such translations; the subconscious performs the workload. Only occasionally perhaps when we search for the right word, the word that matches an idea, are we aware of the structure of thinking. It is likely that the brain separately stores the characteristics of a word: the vocabulary classification, meaning, spelling and sound of words. The vocabulary classification details when the word may be used. For instance, it may indicate that a word is from the French language or it may indicate that a word is restricted by etiquette. This attribute is distinct from the word's meaning. Word meanings are learnt. This is only possible because we start with some innate rudimentary capabilities (eg processing speech into phonemes) and knowledge (eg rules of logic).

Now, most memories are much more complicated. Memories can record facts, associations, meanings and feelings. These memories are built upon other innate subconscious abilities. Thus we have autonomous control systems eg the system designed to regulate body temperature. The system functions by receipt of input signal, by recognizing that this is a core-body temperature, by determining the relevance of the signal (eg acceptable, too low or too high) and choosing a regulatory response signal.

There is no reason to suppose that the autonomous control system initially knows exactly how to choose its response. However, it can learn by experience because another monitoring circuit watches it. Just as some (but not all) birds are pre-wired to enjoy singing specific songs, we may have autonomous approval circuits. If the regulatory error is growing, the monitor will disapprove, if the regulatory error shrinks the monitor will approve. If the error is negligible, then the monitor idles in watch mode.

Note that I am using emotionally judgemental words to describe simple subconscious signals. This use is a subtle clue that such basic drives become blended characteristics, which enrich our consciously experienced emotions. Furthermore this subconscious monitor taps into all the electrochemical indicators that control and signal physical processes. For instance adopt a relaxed posture in the dentists chair and you will become less tense than if you had adopted a protective posture. This is effective whether you believe this technique will work or not.

Does this help answer the mind-body problem? Well in the absence of tenable positions, philosophers have seemed quite happy to hold untenable views. The associations of mind and brain are so strong that mind and matter cannot be the distinctly separate entities. Obviously physical affects are consequences of either physical causes or mental intentions. Hence mind and brain must be related. However, it is pointless to argue (mental monism) that everything could be a mental illusion. The world of my experience is, after all, the only world that I seek to explain. This external world is not pure mind – I genuinely need air, water and food to survive. Neither should we assume that mind is a derivative of matter ie everything is matter. After all, mentality is essentially qualitative whereas matter is essentially quantitative. Computers (hardware plus software) do not have emotions. Computers do not have free will. Computers are rational. This was Descartes' argument against materialism. Naturally Descartes used different terminology and a different language.

Thus the evidence clearly is in favour of mind and matter being formed from something else, something new. This neutral monism is what we shall be arguing. There is a common form, the aether, which is the source of both the mental and the physical. Thus we have seen some of the steps needed for mentality. The mind, like so much else, is a hierarchy. The deepest mystery is the secret of consciousness. This secret explains how mental processes give rise to personal experience.

Empirical Psychology

We will argue that the mind has direct access to the macroscopic aether. We will now ask whether we can get some ideas about the universal langue of the aether from empirical studies of human psychology.

Apparently there are several common psychological theories of personality. Currently the most popular theory is known as the Neo-PI (new personality indicator); this theory is also known by the catchy phrase, the big five. These five psychological traits were inferred after a comprehensive analysis of the English language's vocabulary. It is said that some other languages do not seem to yield the same basis. However, note that the English language is extremely versatile and it has words inherited from a very large number of other languages. Thus if the analysis were only to work for some languages we would suppose English is likely to be amongst these. By this analysis, researches have suggested that the dominant behavioural traits are neuroticism, conscientiousness, introversion, openness and agreeableness. Naturally the behavioural traits are strongly coloured by individual psychological needs. There is evidence that these personality traits are reasonably stable over a person's lifetime. Thus an individual's details are influenced by genetic (perhaps hormonal) and early (or fixed) environmental features. Obviously these typical modes of mental behaviour are only an indirect indicator of the options available for mental processing of information. We shall exclude neuroticism from consideration; neuroticism represents an inconsistent choice of mode of thinking. Hence this leaves us with four dominating modes for mental processing of thoughts. We can refer to these four modes as psychological dimensions.

Now, Myers and Briggs developed an earlier psychological-type indicator. This mother-daughter team based their indicators upon the deliberations of Carl Jung. Jung's analysis was directly focussed on considering the ways in which people access and process information. Thus this Jungian scheme may give a slightly clearer picture of the fundamentals of mental processes in which we are interested. The Myers-Briggs type indicator (MBTI) scheme also has four dominating styles.

In practice, it is straightforward to assess a subject's personality for all such schemes. Hence psychologists have investigated correlations between the results of these personality schemes. Subjects are required to answer a questionnaire. Then, quantitative comparisons are made between the different sets of indicator. Thus we find that two of the MBTI styles are closely correlated to Neo-PI psychological dimensions and the other two MBTI styles are more loosely related to Neo-PI psychological dimensions. The table below lists these approximate correlations.

Table 4: This table shows the approximate correlations between the Neo-PI and the MBTI psychological dimensions.

Neo-PI / MBTI	Conscientious-ness	Intro-version	Open-ness	Agreeable-ness
Perceiving	½			
Intro-version		¾		
Intuitive			¾	
Feeling				½

Both psychological schemes describe aspects of a set of high-level mental modes with subcomponents. However, these schemes do provide a rough starting point for uncovering the structure of thought processes. The schemes are suggesting that the vocabulary of the langue has a hierarchical structure involving quadripartite options.

Before we unveil further details of this mental structure I want to remind you of another quadripartite structure, namely science. Recall that we can talk about scientific observations or the laws of science or the principles of science or the worldview of science. However, no part of science can exist in isolation. Science is not any one of its identified four parts; science is the integrated whole system. This whole of science has four separate states that we can access. Thus this fourfold structure for science is no coincidence. Science describes a fundamental way of analysing the world using thought. I will call any such conceptual unit with exactly four accessible states, a quadron.

As we strip away the sophistication of such a quadron we see simpler and simpler quadronic forms. This uncovering process is not unique. Note that in one (intuitive) view of science we move from information to knowledge, to concept and finally to belief. This quadronic sequence matches observations, laws, principles and worldviews. It also matches the dimensions of the discussed psychological indicators, which are also heavily wrapped in extra semantic overlays. Indeed, in the above table, I took the applied order of psychological dimensions from the order of this quadronic sequence ranging from observation (state 0) to worldview (state 3).

Hence we deduce that when we look closely at human language and at human thought, we observe quadrons such as (information, knowledge, concept, belief) and (get, compare, represent, accept). We might normally regard the verbal states as independent and unrelated concepts; yet they are formally related. Shortly we will show that these sets of words are available options for larger concepts. Thus just as it helps – on occasions – to talk about science, it can be useful to talk about a collective verbal quadron rather than one of its states.

7. The Mental Realm

Canvas

We shall now examine how eventual reality provides an explanation of mind. Note that this hypothesis fills a vacuum; there are no comprehensive rival theories that explain the fundamentals of the mind. Of course there are psychological theories offering views on how the mind works. Furthermore many neurological studies show some details of how the brain functions. However, such approaches have not helped us understand what mind is and what constitutes a mind.

Certainly there has been plenty of idle conjecture about what it takes for matter to display the characteristics of a mind. Some folks think that all you need do to form a mind, is to realize an appropriately organized structure. Some suspect that once matter surpasses a certain level of complexity, a phase with mental awareness can form. Some think that all matter has mentality. Some people think that every flow of information creates conscious awareness. Some argue that mental experience is the handmaiden of causation. Some believe that there could be extreme circumstances in which matter breaks the bounds of scientific determinacy and then becomes completely unpredictable. Somehow this freedom is supposed to equate to consciousness. We can conclude that everyone is guessing; nobody has much of an idea.

Now, let us consider what this new theory has to say. We have noted that matter is simply a description of the statistics of the aether. Hence a definite object in our perceived world is just the outcome of the overwhelmingly certain statistics of the freely exercised opportunities of the aether. Clearly the aether and matter are intricately connected to one another. This connection goes both ways. Changes either to the aether or to matter will almost always influence the other. Note, however, that specifying matter does not specify the aether – only the reverse applies. Potentially the aether has total mastery over matter. Thus Plato's shadow is an acceptable analogy for matter. If the shadow (cf. matter) changes, then light sources or objects (cf. aether) must have changed. If objects move, then the shadow will usually change.

We have noted that the aether is a coherent hierarchical structure of (complex-quaternion) symbolic references. Referrals are qualitative and the details of the aether are representable by language. Thus we recognized that the options of the aether are describable by a language. However, the aethereal field is essentially its description. Consequently the aether is simply the written word expressed in a universal language. We call this universal language, the langue. When the expression of this language is a senseless hullabaloo, then matter tells us everything we could ever need to know about the aether. For the most part, there is no system to impose order on the aethereal charivari. For the most part, matter is an adequate description of reality. However, there are certain self-sustaining aethereal systems that interact with a memory. Memory permits a consistent selection to be made; indeed as memory banks fill, these choices can evolve progressively in time. Memory allows a consistency to the messages expressed in the aether. Consistency leads to rational information. Then the aether may portray a coherent message.

This is how an experience can arise; the aether becomes the canvas for the ideas of the mind. The aether has precisely the right properties. The aether contains rules of grammar and logic. It has a natural vocabulary that can build the richness of experience.

From an aethereal viewpoint matter is a subsidiary concept. We have an aether, which has some clear characteristics that we call matter. The aether has a grammar, a vocabulary and logic. A value in the aether is not the pointed reference of the apeiron. The simplest aethereal value is the expression of a semanteme. A semanteme is the smallest element of an idea that makes sense – the smallest element that can be proved or unproved, true or false. Here we do not think in terms of words or phrases but sentences. Hence a semanteme effectively corresponds to the meaning of an elemental sentence. This semanteme has four parts, which will be called lexemes. These four lexemes are a noun and a verb each with individual logical senses (± signs).

The aethereal quaternion structure provides the vocabulary of the noun and verb. However, we do not attach individual meanings directly to aethereal designators. A quaternion has one scalar and three vector directions. Each vector direction is equivalent. One vector direction has no different significance to another vector direction. Thus our vocabulary is only sensitive to the characteristic number (0 to 3) of aethereal vector dimensions that are displayed in an aethereal semanteme. This is how we generate the quadronic word groupings. Furthermore a hierarchy of aethereal semantemes make a practical aethereal remark.

Let us look at the nounal quadron whilst we keep the verbal quadron on its scalar setting. We observe how a repeated basic identification or selection builds the meaning of a nounal quadron (item, type, rule, conviction). Hence these are some of the primal lexemes of the langue. Note that selection is at the core of the meanings of the langue. The quadron states are a sequence built upon choices. The table shows this quadron format. Furthermore the aether is but a figurative chalkboard on which the langue is writ. We have to tie this writing to matter (by using memory and feedback) to achieve rational sense in an otherwise random scribbling.

Table 5: An illustration of the quadronic format of the vocabulary of a semanteme.

	Abbreviation	Semanteme	Question
0	Selection	Datum-0 is identified	Which one?
1	Classification	Selection is like datum-1	What type?
2	Explanation	Classification is like rule	Why like this?
3	Belief	Explanation is like conviction	How to regard?

Now, language and conscious thought are two related but quite distinct concepts. So how do they arise? Well, it is likely that the benefits of good communication delivered the evolutionary pressure for the development of sophisticated conscious thought. Hence our conscious thoughts are usually transcribed directly into internal silent echoes of spoken words. These internalized expressions of words are in a learnt common language of communication. Thus language is the default; however, we can clothe our thoughts in any combination of internal reconstructions of sensory echoes – as we do in a dream.

In whatever form of our conscious thoughts become aware to us, more basic thought primitives exist in a different independent form. This primitive form is invaluable because it lacks the ambiguities of the English language and may encapsulate the gist of a meaning without using English words. Occasionally in recalling a meaning these thought primitives hang without an immediate conscious translation. Thus as we struggle to identify the correct word, the unstated idea is available only in the pristine unwrapped state – the precursor to conscious thought.

Such thought primitives reside in the aether and follow the rules and vocabulary of the aether. The aether is a discrete indefinitely sized hierarchy fashioned as a tree. This hierarchy shows some resemblance to the grammar of spoken languages. Indeed it is said that all human natural languages obey a common meta-grammar that is the innately acquired link between the aether and the grammar of vocalized speech. This meta-grammar opens the channel for vocabulary and grammar learning by infants. The languages learnt by these children exhibit all of these features of the aether and the meta-grammar. However, spoken language also contains many extraneous features.

Thus let us ignore pronouns (he, she and it) and other determiner implied referrals (eg this and that). Let us ignore the definite and indefinite articles. Then, we observe that paired nouns and verbs build the grammar syntax of a sentence. A trio of subject, verb and object (as in cat ate dog) does not form the simplest sentence. The simplest sentence looks like the duo of subject and verb as in, Newton smiled. In the figure we show how a more complicated sentence is built from duos of verb and noun. Note that the verb, to be, (with prepositions as required) fills otherwise empty entries associated with many qualifiers.

This illustration shows how natural languages resemble the inferred aspects of the langue. However, spoken natural languages have additional non-aethereal communication constraints eg finite acoustic bandwidth and necessary redundancy to cope with a listener's occasional lapses of concentration. These communication constraints fix the sophistication of our languages. This sophistication enriches our language by adding further features. However, these natural languages are not our primary concern. For more details see the works of linguists like Noam Chomsky who first noted the inherent noun-verb structure of spoken language.

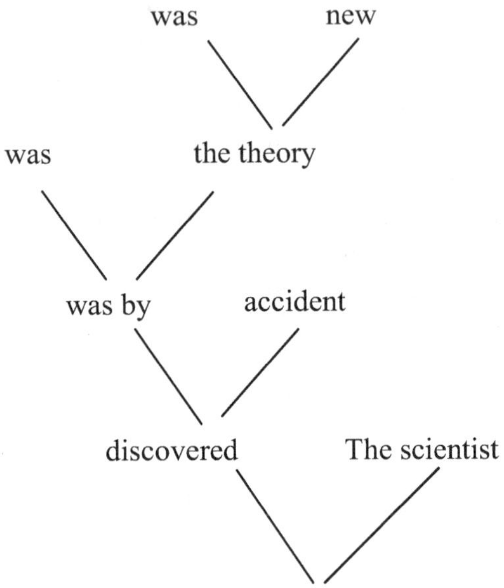

Figure 5: Schematic verb-noun representation of the sentence: The scientist accidentally discovered the new theory.

Introspection

The collective-coordinate treatment delivering the first indication of matter accomplishes a transformation of quaternion-vector space. This transformation changes both the aether and space. Thus it replaces the rational foundation space where microscopically Pythagoras' theorem applies unmodified, by a space where some signs are switched. Hence in this new macroscopic space, the recognizable Lorentzian transformations of special relativity apply locally. Thus this new space both behaves according to the observed laws of physics, and contains matter. Clearly this is our observable space and the geometrical signature of this space shows that it is not fundamental.

The collective analysis picks out the coherent harmonies spanning the foundation coordinate space. This identification of harmony applies equally to the aether and to matter. The mathematical procedure leads to a description of discrete particulate matter. Grassmann numbers, familiarly known as particle creation and annihilation operators, assign the status of particles to the expressions. Likewise the aether also becomes segregated into discrete fundamental units. We shall refer to these discrete aethereal units as aethons. Each aethon is responsible for a particle. These aethons do not reside in space; yet one can image their nowhere existence as a set of invisible eight-paged notebooks (one on each branch) carried around by fundamental particles. In true quantal fashion the idle gossip written on the notebook pages is only reliably stable when the message becomes readable (ie reinforced to classical levels).

Obviously we do not need to discuss matter. However, I will keep referring to allied particle attributes since they are our guide in the uncharted world of aethons. Our knowledge of particles will show us how the aethons assemble into a mind. Like quantum matter, aethons are global although generally we only need consider globalization over the range of interactions.

Holism is a difficult concept for anyone with little knowledge of quantum mechanics. Therefore let me try to explain some of the issues. In the everyday (classical-mechanical) world, every cricket ball manufactured is marginally different from every other cricket ball. Thus we can distinguish Alice's ball from Ben's ball – after all, they are kept in different boxes and Don Bradman has signed Alice's ball. Indeed we label many items with marks and serial or registration numbers. If we have labelled a thing, we have attributed uniqueness to it. However, we do not have to do the labelling; we just need it to be conceivably possible to add the tag.

Now, the hallmark of classical reality is the definiteness of things. Thus it is no surprise that we can label everyday items. However, this labelling becomes impossible when we peer into the less-certain quantum world. There is no practical way to distinguish one electron from another electron. We cannot add a tag to an electron without making it something else. Note that location tells us about multi-particle states not individual particles. Thus suppose there were only two electrons in the universe and these are centred on Alice's ball and Ben's ball. Then, in fact, each electron would be partially centred on each ball. This is the intrinsic holistic or global characteristic exhibited by quantum mechanics. The quick thinkers may be wondering how we can do any experiments when everything is so interconnected. Well if Alice's ball and Ben's ball are very far apart, then we can ignore the interaction between states centred on these balls. Consequently, because particle identity cannot affect any outcome in our space, we can pretend that there are separate electrons around each ball.

Likewise aethons can be portrayed by proxies localized within strongly interacting systems. These aethons specify the similarly localized particles. Hence the finite set of aethereal message states assigns all of a fundamental particle's properties. Aethons are only distinguishable through the messages that they carry.

Now, we may use quite different types of description to report the same things. For instance brain activity is associated with the macroscopic flow of chemical and electrical signals. These flows are currents of matter and hence currents of aethons. Either statistics or energy drives the material flows, depending upon your viewpoint. The macroscopic circuit of flowing and interacting aethons is ordered or in a low-lying energy state – again depending on the mode of description. The ordering promotes macroscopic coherence amongst the aethons – a discernible message is portrayed.

What turns out to be self-monitoring accompanies this message. Such self-monitoring is no more than a verification and feedback loop that echoes an aethereal sentiment. Thus the original message is written in the langue, and to create awareness we need to monitor and translate it. Fortunately the brain is built to provide a montage of feedback opportunities. A feedback loop that reads can reinforce or remove information. This choice is an essential part of mental activity. Thus the brain is a self-programming organ; furthermore it can add more circuits to the feedback loop. In time, these additions become of major importance as we learn human language to power our thought processes. In practice the brain's learning activities involve building and strengthening links. However, the linguistic mapping of languages could be a difficult procedure to master. The infant human brain is primed to learn human languages; yet it still takes a year or so to fathom a limited form of grammar and vocabulary. The purpose of learning a human language is to enable the recording, monitoring and communication of meaning.

Let us switch clearly back to physical-world descriptions for a while. Those who are not deaf can recognize communicated sounds. The brain's internal response is what we interpret as sound. This response signal can be stored in the structures of the brain. We learn to associate these correlates of sound patterns with aethereal meaning. Then, these correlates take on a role of their own as the silent sound of our thoughts.

The self-monitoring circuit spans all brain activities. These activities include regulation of bodily functions, learning/programming activities, processing of sensor stimuli, processing of aethereal meanings and control of body activities. Conscious awareness is no single one of these functions; instead it is their integrated whole. This is where the richness of experience arises. For instance human feelings are primarily an aethereal belief accompanied by an instinctual drive.

The active circuitry of an aethereal-material current assigns the bounds of the mental self. The monitoring of this circuitry confers a holistic awareness of our surroundings, our self, our thoughts and feelings, the passage of time, etc. This is how we form the holistic self.

We form our thoughts in the aether; yet we only appreciate them, after a short processing delay, as a human-language echo relayed by the subconscious. We create the thoughts in our minds by filling in a multiple-choice questionnaire within the aether. Thus these messages and free will are the consequences of exercising a selection. Such freedom of choice is the natural state. After all, the apeiron is an anarchist's dream world. On material storeys, freedom can be consistently or inconsistently applied. For example scientific causal laws of matter are statistical laws of freedom applying when there is no consistent will.

Now you can see the full worldview of eventual reality. Reality affords us a multi-layered view of the universe. Each layer possesses different descriptors and different rules. We call this structure the multi-storey edifice of reality. This edifice has a self-authenticating foundation storey. This foundation is the font of reason justifying existence and rationality. We describe this fundamental reality as a land of qualitative wonder. It is without time or space, matter or motion. It is the essence of being – unrestrained opportunity in the visualized form of an indeterminate network. The basic unit of this foundation is the event, a pointing monad. We construct new representations of this storey (and hence build the higher storeys of the edifice) by creating and expanding a viewing scale. First we find space then matter and then mind.

Mind

For centuries there has been a hard fought philosophical battle raging between dualists and monists. Eventual reality falls into the neutral-monism camp on the foundation storey. However, we do invent a form of dualism on higher storeys. Let us consider why our explanations are so much more convincing that any other alternative.

Descartes was the archetypical dualist. Most dualists believe that there is more to living creatures than their material parts. They believe in some extra *élan vital*, a life force. This immaterial essence breathes life into otherwise inert matter. Naturally this view is often associated with religions. In contrast most monists are materialists. They believe all mysteries of the universe are explainable. We detect, observe and measure, space and matter. Positing anything else is taken as unnecessary and thus is unjustified. Hence life is a yet to be discovered macroscopic emergent phenomenon of matter. Naturally this view is often associated with science.

Most likely you belonged to one of these camps. Can I convince you that both camps are way off beam? Well, their failings are clearly evidenced by the questions that they cannot answer: What systems are conscious? Is mind a consequence of matter? Is all matter conscious? Are all brains conscious? Are minds necessary? Which parts of a person are conscious? How does consciousness interact with matter? What is a thought? Are we more conscious than other animals? Could we become even more conscious? Did there have to be minds in the universe? How and why did evolution suddenly conjure conscious awareness? When do we have our first conscious experience? Is everyone conscious? Can different minds interact directly? Do we have free will? Are minds located within space?

There is one way to answer all these questions. This answer comes with the authority of an *ab initio* hypothesis that explains existence, time, space, matter, science, mathematics, logic and indeed mind. We must turn our thinking around. The edifice of reality is founded upon a bounty of free choices. Thus it is determinism (not free will) that needs to be explained. Storey by storey, the first principle teaches freedom to submit to new rules. Eventually free choice is lost and dim certainty rules the roost. However, in the intermediate storeys we find what is needed for consciousness. We see a duopoly of space and aether. The aether is a tablet waiting for intelligent scribbles. Thus aethereal systems show freedom of expression and rational behaviour. Material systems can form useful memory storage and retrieval functions. Matter and the aether are not distinct. Hence the aether can move matter and can be informed by matter. Functioning systems can develop. Furthermore feedback circuits can regulate all processes including other feedback loops. This complex mix creates the opportunity for and inevitability of mind. Evolution has assembled minds because they are of benefit and they are readily available outgrowths of a responsive nervous system.

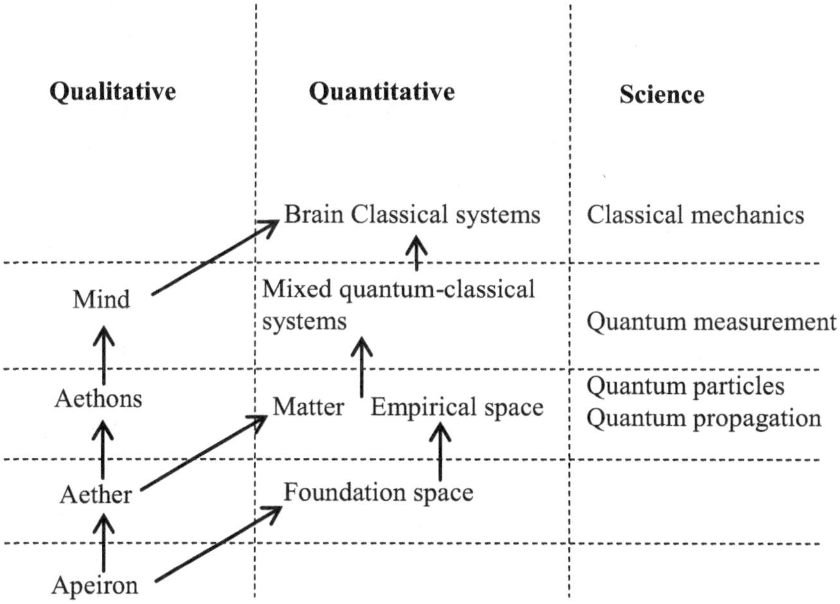

Figure 6: Illustration of the location of mind within the hierarchical storeys of reality.

Hence the mind is a high-storey activity sitting on the storey that mixes classical certainty with the last fling of apeironic freedom. Here we can self-monitor, regulate, encode, translate and reliably accomplish actions. Here we can strike the right balance between being able to control and being able to be controlled. Classical systems are stable and reliable ie dull and boring. Quantum systems retain some underlying flexibilities. However, we must look beyond the matter to the aethons. The aethons are the thought processors.

Awareness comes from the translation of these meanings, perceptions, feelings and associations into brain responses.

The mental system learns and grows. Our aethereal choices actively design routines within the brain. In turn, the data supplied by the brain influence the choices made by the aether. It is all very circular. However, this is how we develop. As neonates, we start with a seed (like a computer BIOS) – a basic life-support operating systems. We grow a catalogue of mental behaviours that suits our environment and inherent temperament. Indeed this is the pattern for most multicellular animals. So why are not we all similarly capable? What makes the difference between a human and a bee or between a dolphin and a gnat?

Certainly we should attempt to answer this question. Apparently most philosophers do not like people attributing human-like emotions or understandings to other animals. They think that we can never appreciate what it is like to be a bee or a fruit bat. The reverse is undoubtedly true. I do not expect a bee to be able to appreciate a human experience. This is because nature has selected us for mental versatility – we possess omni-capable (ie scientific) minds. In contrast bees have either not needed or not been able to support this flexibility. I suspect that many philosophers have simply been reading the wrong books. Perhaps they suspect that white mice are the real scientists and we are just their unsuspecting laboratory specimens!

Anyway, the major life forms start at birth with various sensory support systems. These systems drive sophisticated regulatory systems. Figuratively, as we develop, we encase these systems in layers with further monitoring and control functions. This layering determines capacities for mental functioning. However, each layer imposes a further energy load upon the individual. Thus evolution endeavours to select the optimum functionality level for the creature's niche. For all creatures with a particular functionality, this functionality is substantially the same. Nevertheless it may differ in extent. Since happenstance has richly imbued us (*animans scientifica*) with features of all types, we can indeed imagine what it would be like to have a less capable mind like that of an elephant or a penguin.

How is the mind structured? Well, evolution designed minds. Hence it is unsurprising that they are symbolically complex. Whilst a linear hierarchy is a crude approximation, it can help discuss the functional issues. Machines often have feedback systems. People can program hardware to exhibit flexibility. Systems limited to this basic behaviour have no minds. Thus we begin to talk about minds when there are feedback loops that symbolically represent and monitor other feedback loops. If these incorporate programmed instinctual drives and are aethereal, then they correspond to emotional experiences. Furthermore when we act upon our feelings and monitor our performance in the world, then we begin to construct internal working models of nature and ourselves.

Application	**Comprehension**	**Construct worldviews**
	Explore	**Investigate reason, experience and purpose**
	Communication	**Fluent communication language**
Experience	**Feelings**	**Symbolic experience of the world**
	Symbolism	**Uses stylized symbolic as generic concepts**
	Emotive	**Assess and redirect urges**
Flexibility	**Learning**	**Uses symbolic classifications**
	Adaptive	**Self-programming system, subjugates response**
	Memory	**Access to read/write records**
Feedback	**Monitoring**	**Uses symbolic values for sense data**
	Drive	**Innate response to sense data**
	Perception	**Direct sense data monitoring**

Figure 7: An approximate linear representation of the growth of mind.

Bear in mind, that minds are necessarily aethereal. The aether is the digital hierarchical tablet for mentality. The aether forms a hierarchy of semantemes. Each semanteme is a blend of meanings, grammar and logic. Meanings grow from the simple referential beginnings. They are always accompanied by an innate logic.

Technologies

This exploration of the mind is still developing. Clearly a full understanding of the mind will herald in a new era of opportunity for new technologies. Occasionally the implications of some technologies have led some people to wish that the scientists had not done the basic research. In defence of science, supporters often say that scientific results are neither good nor bad – whereas society's use of the science may be dangerous. However, I would like to be contrary. Freely available scientific knowledge can only be beneficial although sometimes the timing of its arrival can be poor (more often for the messenger). Any ensuing oppression of science arises because of other non-scientific problems. To deny that useful knowledge is inherently good is to treat humanity as immature and incompetent. Surely to regard our fellow humans as stupid and deliberately to keep them ignorant has to be the wrong option.

This does not mean that there should be a free for all in the exploitation of science or in the methods used to advance science. It means that a fully educated and informed populous must discuss and decide what will be allowed. Thus everyone deserves the best available knowledge before a consensus is forged.

Some mind-technology near-term predictions are easy to make. For clarity let me note the following usage of words computer, robot and cyborg. Computers predictably follow algorithmic solutions; they can communicate and exercise primitive controls. Their role is at the dumb end of the automated response. Hence computers are said to be mindless and brainless. Robots, such as those that replace humans on assembly lines, can recognize and respond in a more sophisticated way (than computers) to their environments. Hence robots are described as having electronic brains. Robots (including human-looking androids) perform various designated functions. They have brains but they are mindless. Note that we are not letting our language presume a Cartesian split between mind and brain. We are simply allowing a broader definition of brain. All minds are associated with a brain, not all machine brains have an associated mind.

However, if we gave awareness and intelligence to a machine genuine (ie a mind), then the machine would be called a cybernetic life form (cyborg). Note that we also use the term cyborg to refer to intelligent hybrids of technology and organic life forms. Now, minor human enhancements (not worthy of the cyborg label) like the cochlea ear implant are already here. This medical technology offers a lot of promise; however, it also shows that our ability to interface the brain with manufactured devices is currently exceedingly limited. Thus we have plenty of time to come to terms with the widespread implications of eventual reality.

The exploitation of such mind-machine technologies will affect the cultural, political, religious as well as the technological aspects of societies. In time the notions of what it means to be alive and to be human may change. Eventual reality celebrates your free will, your ability to contribute to decisions about these changes. To participate in these changes, you must think on the issues!

Properly regulated technology has substantially benefited mankind. Intelligent machines offer unprecedented new opportunities and new benefits. Indeed we can seek advice from intelligent technology as to how to regulate intelligent technology. We do not have to give cyborgs jobs that we want to do! However, there are plenty of monotonous or dangerous tasks.

Please forgive me if I state the obvious. If we create minds that are more intelligent than we are, then we can be assured that they will be less stupid than we are. Possibilities for our future are almost endless. We will live longer and healthier lives. Any job that we are not willing or able to perform can be given to a cyborg. Imagine sending cyborgs, capable of rearing the first generation of a new civilization of humans, to another galaxy with human eggs and sperm. Suppose we want to improve the design of a fusion (ie clean nuclear) reactor. Suppose we want to regulate the world economy. Suppose we wish to build a space station on the Moon. Suppose we wish to catalogue life on Earth. Suppose we wish to mine the asteroids for minerals. The unity of humanity could be defined by our sameness contrasted against our difference from artificial life.

However, I think that I should warn you that these possibilities are likely to be dwarfed by others yet unimagined. Perhaps a new age is about to dawn – no one can really know. The human story is one of a progression of views from Dawn-Age mythology to Steam-Age classical science to Computer-Age quantum reality to the as-yet unimaginable vista of eventual reality. Would not it be wonderful if this came to pass? So let us consider how our society is coping with science. Do we need to do things better?

8. Implications for Society

Heraclitus argued that the universe was a relentless non-winnable contest between uncompromising opposites: change and stasis, hot and cold, order and chaos, everything and nothing. Yet we have found these pairs of opposites to be completed duets. The universe reveals a conflated unity of cooperative concepts that nestle together across the storeys of reality. Each side of a pair of opposites needs the other to acquire relevance. No side is exclusive and each begets the other like life and death.

The key lesson of eventual reality is that the universe is an unconstrained mess of glorious choice and that each particular description holds the seeds of its opposite. This is not a recipe for despair or even amorality, but it is a recipe for constructive achievement. The entire universe is nothing but a stacked hierarchy of fated cooperation. Clearly decisions on each storey of reality do matter. Yet they are the celebratory bounty of the interminable indecisiveness of the apeiron. Inanimate matter vacillates. The role of animated matter is to apply choice meaningfully. Intelligent life's noble purpose is to share stewardship of the material realm, to act as wise guardians of the entire observed universe.

Thus applied wisdom is the goal for individuals, communities, societies, species and planetary ecosystems. Achieving individual responsibility ought to be easy; achieving societal responsibility is difficult. I do not have any certain answers only opinions about answers.

There is a further reason why this final chapter discusses aspects of the interactions between science and societal powerbrokers. If we do not learn the lessons of the past, history will repeat.

We live in an increasingly unstable world. For several reasons the world's geopolitical map will probably change dramatically over the next quarter century. For half a millennium there have always been some Western societies, which have strongly backed progress in modern science. However, we cannot rely upon this necessary support continuing. Indeed it has not always been so. Remember the Dark Ages. At the end of the Dark Ages, Europe had less knowledge of mathematics and philosophy than Plato had. Thus let us look at some parallels with events in history.

Once, politicians put Socrates on trial for his unsavoury views. Today, some Western politicians choose to dispute any unwelcome science. True, it is much safer being a scientist than being a politician – however, I want to draw attention to indications of consequences for science not for people.

Once, Phillip the Macedonian brought unity to Greece. Then a consensus grew into cultural strictures upon free thought. A new productive age of philosophy dawned, an age that was unresponsive to notions of fundamental reality. Today European borders are vanishing and European governments are coordinating policies. Do not misinterpret my message; I am a supporter of European union, but we cannot ignore the lessons of history. We must ensure a balance of commonality and independence, one that guards against an imposed rigidity in thinking through, say, uniformity of teaching curricula.

Once, a strong Roman Empire emerged in Europe. The Greeks lost mastery of their future, and national pride turned to stories of bygone achievements. Rome greatly admired Greek philosophy – yet the Romans were just too practical, they could not preserve the intangible progress in fundamental philosophical achievements. Today, the United States shows signs of looking to the past rather than the future. The United States has spent much of its previously earned political capital; it is heavily financially indebted and other nations are emerging as new powerful political and economic powers. These global changes in geopolitical power are inevitable.

Maybe, if we are aware of the pitfalls for scientific progress, then we can sidestep them. We form an unhealthy chimera when we graft pure (unpredictable) science onto cultures that have a set agenda, that comprehensively regulate and control, that measure performance. This is somewhat ironic as science has an agenda to regulate, control and measure performance.

Sophistry and Science

What is truth? We live in a time when many insincere people regard truth as a commodity to be manufactured, sold, bought and owned. Politicians gleefully distort the truth to suit their aims and ambitions. Pressure groups invent or exaggerate and assert their own versions of truths in order to exert an undue influence over public opinion. The media either trivialize or dramatize the truth to sell 'copy' and thereby increase their advertising revenue. Then, many social groupings can require adherence to a pre-packaged constructed truth as a condition of acceptance into the group. Our world is certainly more complex today than the agrarian and nomadic worlds of our ancestors.

What should anyone (without the necessary detailed background knowledge) believe – when there is always someone who is prepared, for personal benefit, to present the most untenable contrary argument? Is the Pope a Roman Catholic? Is the Earth flat? I hope that you answered: Yes then No to these questions. However, anyone can be contrary. If I wanted to start an argument I could say that the Pope is not Roman and the Earth's surface is indeed locally flat everywhere. Alternatively I could maintain that the two questions are nonsense: Any church with a leader cannot be a catholic church. There is no unique surface for the Earth. Clearly it is too easy to spout contentious irrelevance!

Therefore let us first try to understand some of the factors that have led to a regard for the truth becoming so degraded and soiled. In Western democracies a political class mould the laws of the land; a substantial proportion of this body were trained as lawyers. Now, the adversarial system of law is dominant but not universal. Thus most lawyers are taught that the truth is for the courts to decide, their legal function is to put a convincing case (no matter how flimsy the evidence) in support of their assigned role. It is unsurprising that over time our politicians come to regard this adversarial model as appropriate for all reaches of society and they imagine that truth should only be decided by persuasive argument and never by demonstrative fact.

The political acceptance of a flexible truth debases honour, trust, integrity and honesty. We should not blame the politicians; this outcome is of our making, we have consistently elected such an unrepresentative bunch of lawmakers. I am not advocating that these politicians be cast aside and replaced by those like scientists trained in analysis and reason. A dominance of scientists (adherents of the facts) would undoubtedly have created an even larger array of quite different problems.

However, so many of today's critical problems are associated with an ignorant misuse or abuse of scientific knowledge. Consider some of the pressing issues: global warming and climate change; HIV/AIDS; SARS and mad-cow disease; other transgenic diseases like avian influenza; environmental pollution by plastics, pesticides and other endocrine disrupting chemicals and by transport vehicles and heavy industry; the alarming rate of extinction of species; recycling and disposal dilemmas; planning power generation systems; moderating and shifting consumption as Earth's natural resources (mineral and biological) are depleted; transportation and biological quarantine procedures, etc.

Before anyone becomes too alarmed, let me state that, like most scientists, I am an optimist. There are always plenty of excellent solutions to be devised; however, most likely the best solutions will take many years (20+) to design and implement. Scientists need plenty of time to gather all the facts and construct practical models of appropriate subsystems. Thus scientific research work must begin when we first identify potential problems. These problems should not be left until the consequences become so severe that governments can glorify in their strong leadership during a crisis.

There are many reasons why warnings from the scientific community are constantly ignored. Politicians have a short term agenda, the stitch-in-time argument just does not work. However, the biggest problem is one of miscommunication. Scientists are human and many get carried away occasionally. However, scientific integrity prevents us from deliberately gilding the lily. These scientific problems are all highly complex, our operational tools are never as big as our appetites for knowledge.

Consider the prediction of the weather. We all protest when the weather forecasters occasionally get their predictions substantially wrong. However, the reality is that all predictions come with a level of uncertainty. The scientists who compute the forecast will not claim certainty. This hedging-of-bets is common across all of science, all scientific claims are qualified. In politics the opposite is true. Politicians pretend to know what to do with absolute certainty. Furthermore a political issue and its resolution have to be 'sound-bite' simple. Opponents accuse a politician who (from the outset) proposes a wishy-washy balanced approach, of being indecisive and losing his/her nerve for a good fight. Such a statesman is said to have no 'bottle' and hence supposedly no leadership abilities, and is quickly relegated to the backbenches. This crass destructive internal dynamic sets the political agenda: good intentions, greed, ridicule and subterfuge.

Politicians desperately want to be told that something will definitely happen if they do not do anything – this way they have a scapegoat. Scientific integrity often prevents scientists from making such definitive statements. Anyone who was to stray beyond a legitimate scientific assertion would be howled down by his or her colleagues in the scientific community. The scientific community preserves and expands scientific knowledge and understanding. It is important for us, as individuals, to be clear and precise, and not to paper over gaps in knowledge and overstate what we know or believe. Some ignoramuses view this trait as cowardice. In the 1980s I can recall being told that politicians referred to the scientific community as wimps because we did not take to the streets in protest about their science policies. Therefore general funding declined and societal problems multiplied. However, very little riles the scientific community; certainly being called wimps (by jerks) does not get us upset. Anyway our real concerns tend to be elsewhere. Most politicians have yet to wake up to the fact that, in the modern world, science has become much more relevant to their decision processes.

We all know that smoking kills (costing the community dearly in emotional and financial ways), but there are potentially individuals who can lead full lives without any smoking related illnesses – these individuals are the so-called *exceptions that prove the rule*. The scientific evidence linking smoking with many serious illnesses (cancer, heart disease and stroke) was clear for many decades before governments condescended to take any action. There is comfort in companionship of other inattentive governments. Could it be that national governments felt that they could delay and wait for another national jurisdiction to take the lead? Could there have been a shortage of leadership qualities in our governments? Could our assertive decisive leaders have no 'bottle' after all?

Anyway to support the delay (and collect all those tax dollars) governments may point unjustifiably to slick marketing presentations and maverick scientific opinion. Maverick scientific opinion is an essential ingredient of the inner workings of science; it ensures the continual testing and challenging of the veracity of all scientific information. However, everyone should recognize these nonconformist hypotheses as outrider views; they do not hold the same status as the strongly supported mainstream views. There is a parallel with events in the natural world. A young male lion may contest for dominance of a pride; a challenge per se does not confer an equal chance of success to the challenger. Indeed I suspect that the inexperienced juvenile lion is likely to lose his first battle. The more times a male lion has successfully defended his pride, the greater our expectation (barring injury and infirmity) that he is strong and aggressive and thus he will continue to be successful. Likewise the continual success of our mainstream scientific views gives us a strong confidence in their explanations and predictions.

Setting Priorities

Practicing scientists see pure science as an endeavour to uncover the truths of nature for the benefit of everyone. Inevitably very little science can be self-funded by scientists. Overwhelmingly governments control scientists and science policy. Many large projects with photo opportunities for politicians or with nationalistic agenda (eg the space race or promises of Nobel prizes for its citizens) are given the go-ahead. These projects are politically akin to the government funding of sports' stadia, sports' institutes, national parks, museums, national orchestras, concert halls or major festivals and events such as a grand-prix motor race or an Olympics. The publicity gives the impression that governments are active in supporting science.

It is surprising that governments spend quite small sums of money on bread-and-butter pure science. The outcomes of this forward-looking science are quite unpredictable. However, it is this science that potentially

- supports future industrial development,
- protects (through understanding) society from catastrophes (asteroids, tsunamis, volcanism, epidemics, climate change, magnetic-pole reversal) and

- helps us to understand our place in the natural world (the Earth is not flat; the Sun is not the centre of the universe; we are terrestrial animals, the product of evolution and sharing a planet with countless other life forms).

Science folklore says that Queen Victoria once visited the renowned experimental physicist, Michael Faraday, in his laboratory. (Another version of this folk tale has an unknown lady visiting Thomas Edison.) Faraday patiently demonstrated some of the curiosities of electricity and magnetism to her. After the demonstration the Queen was bemused and is reputed to have asked: *But what is it useful for?* Slightly miffed, Michael Faraday retorted: *Ma'am, what use is a baby.* It is one of the wonders of science that we often do not know how things will turn out. Note that a baby is not simply a DNA replication vehicle. Some babies will grow up to deliver great achievements to a society or even mankind. However, all babies have talents hidden from view; they are valued for themselves irrespective of any future greatness. At birth, babies are a joyous bundle of possibilities – like science and indeed the universe.

Today the technological benefits of electricity are obvious to us; very few artefacts (throughout developed societies) either do not use electromagnetism or were not produced by some form of electromagnetic device. Faraday could not predict any of these consequences. The benefits of most science are less dramatic but the gains have no use-by date. A tenth of a cent every day from tomorrow until eternity is a lot of money. Thus as the future enfolds, the economic spin-offs eventually will pay for all the contributing science, over and over again. This is not the reason why scientists become scientists. Science is a vocation. However, it helps (when society is paying you to play) to believe that your work is guaranteed to pay for itself eventually.

Now, in the political sphere, an ignorant public can be cynically manipulated by appeals to patriotism, fear of the uncertain and greed. In this way, often at election times, emotional entreaties are used to gazump rational debate. Fortunately there is an easily applied antidote to inappropriate emotional appeals – humour. Humour displaces other emotions allowing reason to enter a debate. Sadly opposing politicians prefer to rely on voter fatigue rather than start a laughter war. Of course there is also a permanent cure – education.

Educational Resources

We live in a highly organized and complex society. There are many groups with a diverse range of opinions. Political success is judged by the wielding of power. Unfortunately if politicians are continually publicly challenged by adverse groups, then politicians cannot govern by consensus and they must resort to governing by a general acceptance of their authority. There are many ways for governments to strive to obtain such a placid population. In some countries the opposition is ruthlessly silenced. In other countries governments enact laws or regulation to enshrine censorship or bias in the media. The subtle promotion of bias is commonly favoured in many Western democracies.

Commercial media are very sensitive to the wishes of their advertisers. Big business likes stability (to be able to make 20-year investment decisions) and is interested in many future government policies. Naturally big business harbours good relationships with governments. Laws that unnecessarily protect the status quo in media, promote a system that enshrines a fixed historical bias.

Another approach is to require public broadcasters to give equal emphasis and opportunity to alternative views, no matter how wacky these views might be. This technique results in a form of desensitizing and de-educating the public. Gradually the public loses a degree of confidence in the knowledge they have acquired through limited educational study or life experience. Too much in our complicated world seems uncertain and confusing; hence too many adults become voter fodder for those politicians with the regularly repeated over-simplistic message. The politician's message of salvation is clear: *Don't you worry about that; support me and all your problems will go away.*

Simple messages are very attractive to politicians. However, they can be dangerous. The *kiss of death* policy holds one of the simplest and strongest of all messages. This policy proclaims support for an objective, which in the short term is surely achievable; however, the policy is self-defeating and ultimately brings about the opposite outcome. There is a useful parallel with Napoleon Bonaparte's quest to become Emperor of Europe. Having subdued most of Europe, in 1812 Napoleon led a huge battle-ready army of almost 700,000 men into Russia. He succeeded admirably in his objective of quickly capturing Moscow. However, unexpectedly the Russians did not sue for surrender. Napoleon had not destroyed the Russian armies. In fact, as the campaign continued, the Russian army was significantly expanding with trained recruits from far afield. Thus the extended French position became progressively unsustainable. The Russians forced the French into retreat across fields ravished during the earlier fighting. Napoleon's reputation as an invincible commander was ruined; less than 250,000 soldiers from his grand army survived the campaign.

Napoleon was not a fool; he was a great strategic thinker; he was in his element when he planned the Russian campaign. Why did he fail? Well, experts are human too. Napoleon's error lay in the mindset that he had developed earlier. He saw Russia as a larger version of the other European countries that he had been sequentially toppling. Nevertheless his opponents did not share this mindset. Napoleon practically forced the Russians to use a strength that Napoleon had seen as a weakness – the Russian ability to build a powerful army slowly. In these remarks I do not wish to belittle or overlook the Russian determination to succeed at all costs. Both sides paid a terrible price for Napoleon's vanity.

The issue of inflexible mindsets is still pertinent today; the prevalence of political mindsets helps to explain why intelligent people, elected to positions of power, can often make such poor decisions. The parliamentary party room and other similar institutions are where group mindsets are shaped and preserved. Whenever these self-imposed mindsets become too rigid, they start to resemble the mentality of the dark-ages. Freethinkers must keep their thoughts to themselves or sit ineffectively on the cross-benches.

Now, there are also rigid scientific mindsets tied to the paradigms of science. However, science is effective because these scientific paradigms include rules for their abandonment! It is one of our mindset beliefs that science is almost inevitably right but science can be wrong occasionally. Science's *modus operandi* is not to hide potential failings; science actively seeks them out and studies them, hoping to find a reason for modifying the worldview.

It would be a mistake for you to view these remarks as a tirade directed either against politicians or against the ills of a society that is becoming less friendly to science. Elected and unelected politicians only exercise a delegated power – power lies with the people. If politicians scorn science, then it is regrettably because scientists have inadvertently neglected their own responsibility to educate the public about science. There is no shirking of responsibility by any grouping.

Now, it is no secret that the majority of scientists are introverts. However, this does help to explain why scientists do not generally seek public exposure to explain their truly remarkable achievements. Interacting with people invigorates extroverts; this is not true of introverts. Scientists become so engaged in the in-depth detail of their own research that they are unlikely to attempt to explain their work to someone too far outside their own narrow field. Of course this need not be the case. I recall that the Irish mathematician William Hamilton successfully devoted many years of his life to developing a generalized complex algebra (an important constituent of the hypothesis of eventual reality). Hamilton explained the core issue to his young children and was charmingly faced each morning over breakfast by a chorus of young voices asking 'Have you solved the triple product yet, Daddy?'

A daily TV news programme will probably have several specialized sections including a business section, a sports section and a weather section. I am not sure that there will ever be a regular science section on the TV news. Nevertheless the scientific community could aim to achieve this goal. Clearly there is already considerable potential copy material available: many newspapers have at least a weekly science report and there are numerous science hardcopy magazines and somewhat more dubious regular TV and radio series about science.

However, many program controllers dismiss much scientific progress as incomprehensible to the general public (well to arts graduates like themselves)! Nevertheless well-presented science is comprehensible to most members of the general public (possibly barring some of that small minority who become program controllers). (Note that any program controllers who read this book are obviously not amongst that unfortunate group of controllers.) Thus undoubtedly it should be the responsibility of the senior members of the scientific community – those with telegenic personalities, spare time and little to lose if governments and other bodies are annoyed – to reinterpret scientific advances, to promulgate scientific knowledge and to educate the public at large in scientific matters.

It is critical for all of us that advanced societies actively support science. There will not be another Renaissance. There is a symbiotic dependency between science and modern society, the suppression of science is no longer an undesirable option – it is a non-viable option.

We need science to tell us what is happening around us. We need science to correct the excesses caused by the actions of previous generations. We need science to maintain current jobs and economies. We need science to protect us from nature's casual catastrophes. We need science to open up new vistas of undreamt opportunities. We know that the long-term benefits of science are always magnificently unpredictable. When the first electronic computer (with stored programs) was completed in 1948, no one could have envisaged the widespread proliferation of computers that we experience today or any of the myriad of devices equipped with computer chips.

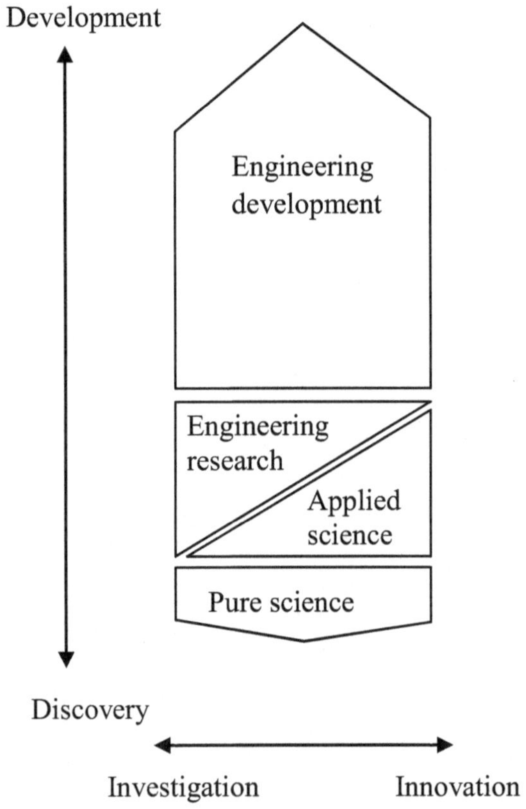

Figure 8: Idealized illustration of the relationship and boundaries between engineering and science.

However, perhaps economic development is not the reason we do any science. We are not the robot slaves that economic rationalists like to imagine. Science is an inspiration. The reason some societies successfully support science and other societies fear science, is cultural. Science brings an understanding of the universe, carefully built brick by brick. Thus science brings unpredictable change. The mysteries of science are worthy challenges to our intellects. Many Earthly creatures use tools and language; science is what makes our species unique on our planet. This is why we need more scientists and more knowledge of science in the community.

Of course I have been grossly unfair in not acknowledging the leading role of engineers in delivering these scientific benefits to society. Engineers are the practitioners responsible for much of the application of scientific knowledge and understanding. On the biological side of science the engineering disciplines are known as medical science, veterinary science, etc. Clearly engineers can rightly claim that they delivered the modern scientific miracles, with science *only* contributing understanding and theory. Naturally we each have a different perspective on what is more important.

I suspect that some readers may not have a clear idea of the difference between an engineer and a scientist. In ultra-simple terms, science seeks explanations whilst engineering has a primary goal of exploitation. However, practice often blurs the distinctions close to the various boundaries. The figure shows where the science-engineering boundary lies. Having already given such a lengthy explanation of science, I will not give a more formal definition of engineering.

Now, engineers have so many opportunities to exploit a glut of earlier scientific discovery, so that sometimes they might feel they cannot keep up with the opportunities afforded by science. However, this abundance does not justify a slowing of the pace of scientific progress. Science does not advance uniformly across a vast front of knowledge. Scientists are attracted to areas that are important or intellectually challenging. However, each decade or so, society reassigns the measure of importance. We need to change these priorities for many reasons eg scientific or technological advances, industrial shortfalls, resource availabilities and natural catastrophes. Scientists are few in number and constantly build upon the old science, as they generate the new science needed. Thus there is a good justification for continuing support for applied science. Does this mean we can ease up on the pure science? Well, it is true that pure science is more remote from the consumer goods and services that our societies appear to value. However, to be effective, applied science needs pure science – just as engineering needs applied science. Furthermore there are also other spectacular rewards that surprisingly follow from the improved understandings of nature.

Recall that Faraday (the experimentalist) and Maxwell (the theoretician) were pure scientists who studied the secrets of electricity and magnetism. Others (like Thomas Edison) turned this knowledge into practical devices. Yet the biggest surprise was the understanding that light was electromagnetic and corresponded to just a small frequency band within the electromagnetic spectrum. Thus it became inevitable that we would invent medical imaging machines, radio telescopes, wireless communication, radar, infrared speed cameras, etc. When pure science completes any small task there are always large unanticipated payoffs on top of the expected payoffs. This is why science is a startling journey of discovery.

Let me give another example directed towards a topical problem. Suppose we want cheaper and more efficient solar cells. Who do we turn to, engineers or scientists? Of course the answer is both. However, we must turn to the scientists first. Science needs to analyse the problem and create new opportunities before they can be exploited.

Thus we can draw a parallel with the family (engineering) that brings a new child into the world with a donated egg and sperm (science). Emotionally the offspring belongs to the family; however, the child would not be possible without the donated gametes (science). This is a partnership; hence the rivalry between science and engineering only extends to claims of importance and favouritism. Indeed experience shows that the borders between the two areas are never exactly clear.

Unfortunately thanks to a failure of education some scientific theories may seem exotic and far-removed from everyday experience. Such theories are an easy target for sophistry. However, these theories are extremely relevant to everyday devices. The Global Positioning System (GPS) could not work usefully without the theory of relativity; we would never have conceived of most of our electronic devises without the theory of quantum mechanics, etc. Sophisticated scientific equipment relies on such physical theories. So much of our immediate world and our being is simple physical matter. However, the associated well-established scientific understanding of matter (ie quantum mechanics) is not even taught in secondary school. Disastrously most scientists will not even learn about modern theories of matter at university. Hence many non-physical scientists (psychologists, physiologists, botanists, biologists, etc) are held back in their work by blissful ignorance of what ought to be common scientific knowledge.

Now, eventual reality is one-step further removed from everyday reality than these physical theories are. We can try guessing; however, we do not yet know what surprises eventual reality will bring to our world. We only know that we have not seen anything yet! Generally the more a theory is fundamental, the more powerful will be the theory's implications for society. Tantalizingly eventual reality promises to unleash unimaginable opportunities for mankind. Rejoice. It is a privilege to be living at the dawn of the first century in which these pickings will literally come to life. Welcome to the real world!

Failure of Tribalism

The supreme authority of the nation state within its own borders is a clear form of extreme authoritarianism. However, in today's world such extremism is no longer a reasonable option. It is possible to travel to any corner of the Earth in just one day. Think back to the founding of most nation states; then it took days to travel to the outer reaches of the nation. In our mobile multi-cultural world, how can a line on a geographical map be anything more than a convenience? Tribal survival dictates that authorities must be accommodating to other tribes and to all individuals.

Thus the independent nation state is an outdated concept. So too is the independent religious tribal congregation. War, hunger, poverty, oppression and selfish luxury are the evils of tribalism. We should remember that people do not cause wars; tribes and their leadership cause wars. A tribe's greed, envy, prejudice and pride can have many global consequences: wars, economic depressions, disease epidemics, depletion of resources and even sea-level changes.

It could be argued that false national pride and business groups have created much of the confusion that (in parts of the industrialized world) has blinded some politicians over many clear issues such as climate change or exploitation of whales in Pacific waters. As this section illustrates, it is easy to find some aspect of everything to be critical about. It is unfortunate that genuine facts are easily criticized and dismissed when those facts conflict with long-established aims and desires.

Just as nation states will need to surrender some powers to a central planetary government; so too might industries establish their own international regulative councils to manage a level playing field and thereby to ensure that good businesses are not out competed by scurrilous commercial concerns engaged in pollution, inadequate worker health and safety etc. Of course global industry could wait for a vindictive world government to impose less benign controls upon them.

Now, it is likely that the subjugation of tribes will mean that we will need fewer standard politicians. This may not seem obvious at first. After all, we still need tiered administrative systems of government. However, although politicians may run an administration they do not administer services. Public servants are responsible for administrative matters. Politicians run the governing body that directs the administrators. Politicians are directors and decision makers. In contrast public servants interpret and implement the decisions of politicians. Politicians have a difficult job for which we continually criticize them. They have to placate lobbyists for society's power brokers, persuade their political colleagues to support them and protect us from thc machinations of politicians of other tribes. Generally (but not always) proficient politicians are bad at the first task, average at the second and good at the third. The reason for this ranking is clear; the political mind understands other politicians. Therefore with fewer politicians, we will need fewer politicians.

Suppose two men arrive at your neighbour's door selling some new technology. The first man is a scientist; the second salesman is a conman who say that he is a scientist. The conman is skilled in the art of persuasion and he is selling junk. The scientist is highly knowledgeable and he is selling valuable merchandise. Now, we know that you would make the correct choice. Would you neighbour? Your neighbour is a politician with limited secondary-school education in science. If your neighbour buys a shonky appliance, the world does not stop. However, what happen if you neighbour lives in the White House, Tiananmen Square, the Kremlin or the Élysée Palace?

This is not a new problem; incessant bad judgements worried Socrates. Socrates taught that a city-state's decision makers should understand the issues before they even seek a position of administrative authority. Socrates constantly reiterated this theme. He would sometimes instruct by articulating a string of analogies. Thus he asked a series of straightforward questions: Whom would you ask to mend a shoe? Whom would you ask to mend a ship? These repairs were the work of skilled trades and the answers were clear-cut. Then Socrates would ask ‘Who should fix the ills of the state?’ This series of questions was a hallmark of Socrates’ teachings of aristocrats. It did not always earn him favour or respect amongst the powerful elite.

Constructive political decision-making has proven harder for today’s politicians. Primarily politicians do not set objectives they mandate solutions. They do not seek consensus; they seek unattainable quick solutions. In parliamentary democracies party leaders are fortunate to acquire an apprenticeship in the art of national governance. However, this experience cannot equip them with the knowledge they need to make sound judgements about technical issues. Politicians routinely ignore the expert scientific advice of their departments even after they have installed other politicians to head these departments and thereby to moderate the advice provided.

Global leaders will never be able to prescribe solutions. They will have to restrict themselves to the simpler task of setting and monitoring goals. You do not have to be a skilled technocrat to accomplish this feat. Indeed our hierarchical minds adopt this same easy-going strategy. Hopefully regional competition will lead lesser administrations to follow this example.

Finally let us remember the legacy of Canute, the one-time King of England, Denmark and Norway. We would be two-faced to abuse King Canute for empirically testing his authority over the tides. Presumably he did not dispute expert scientific advice, just commonsense. In contrast in recent history in different countries our politicians have derided the teachings of science. Thus some politicians have taken it upon themselves to decide the cause of AIDS, the cause of climate change and the validity of the theory of evolution. These views have certainly influenced policies even though legislation may not explicitly enshrine their stated beliefs. Furthermore the twentieth-century history of science in the Soviet Union was also disastrous for many scientists. Political doctrine imposed rigid viewpoints on materialism, evolution, genetics, quantum mechanics, relativity, cosmology, cybernetics, psychology and psychophysiology.

From a scientific perspective, science is not political. To a scientist it almost seems crazy that any politician should ever feel intimidated by science. Yet we overlook the simple fact that science is the most powerful thing in the universe. The power of a black hole, an asteroid, a supernova or an epidemic is resistible. Science tells us how to tame the forces of nature. Indeed science is what makes us, as a species, unique in the history of life on Earth.

Let it be remembered that political power is transient. History is written with a pen not a sword. Everyone remembers Galileo and his ideas. However, how many remember which Pope condemned Galileo to be shunned as a heretic for three hundred and fifty years. We remember the famous Greek philosophers but who recalls which Emperor banished their philosophical teachings? Who can name the monk who isolated Friar Roger Bacon (Francis Bacon's ancestor) because Bacon was dangerously overeducated in mathematics, philosophy and linguistics? Who remembers the Cardinal who condemned Giordano Bruno (1548-1600) for his Copernican and other heretical leanings? It is only recently that the Vatican has expressed sorrow and regret that Bruno was tied naked to a stake and burnt alive. It is only because of the recentness of other tragic events, that we might remember Einstein fled Hitler's reach of influence. Many of Hitler's accomplishments are undone, Einstein's scientific legacy will never be. Science can only build upon them. Thus Hitler's dark name will be forgotten long before Einstein's name.

EPILOGUE

How can I claim to have a solution to what everyone reckons must be an insolvable problem? Certainly many casual thinkers believe that, whereas one should ignore natural-philosophical conclusions because philosophical ideas are not testable, science is fundamentally restricted in its applicability because its laws must be refutable. Hence there are likely to be things that we can never know. If this thinking sounds like mysticism, then it is! Supposedly consciousness, free will and universal existence are prime contenders for this category of hopeless causes. Such unwarranted defeatism is born out of despair. Some tasks seem to get harder. Science encompasses so much more today than ever before. Successful academics are mostly specialists working away on the detail in their specialized discipline. How can anyone successfully wrestle the broader questions? Ah yes, the pace of modern life is a curse.

Despondency creates its own imaginary chains. If you are a sufferer, let me lift these chains for you. Certainly science has a strong focus on experimentally investigable propositions. In contrast, metaphysical philosophy has a strong focus on rational beliefs only indirectly associated with observation. Hence, for instance, philosophers have endlessly debated the mind-body problem without resolution. Yet both science and metaphysics are complete quadrons. Admittedly our natural approaches to these disciplines tend to be from different directions. Nevertheless we cannot overlook the other quadronic components. Now, the discriminator of beliefs is beauty, and a practical comparison of two less-than-beautiful beliefs is often controversial. Hence science (relying on an objective empirical discriminator) powers ahead whilst metaphysics languishes in a quagmire of subjective accusation and counteraccusation. However, we should remember the caution written in fine print: past performance is no guarantee of future performance.

Only perfection makes possible an escape from this metaphysical predicament – the comparison of different forms of beauty. Our beliefs must indisputably match the fundamental rational archetype for all beliefs. Then, the foundation worldview becomes undeniable. This is what we accomplished in our derivation of eventual reality – self-justification of everything from logic to the universe. The beauty of perfection arises from the absence of any underlying beliefs. This is one occasion when bland is blissfully beautiful.

Wait! Surely there is still a problem. Does not Gödel's theorem rule out the possibility of a final answer? Well Gödel showed that some valid statements have to remain unproven in a logical system based upon a finite number of founding premises. More precisely he showed that mathematics (defined by any explicit finite set of axioms) is necessarily either incomplete or inconsistent. Freeman Dyson, Stephen Hawking and others have reshaped Gödel's argument so that it would plausibly apply to a final theory of science. Thus Hawking relied upon the mathematical nature of science. He replaced the logical mathematical axioms by a proposed finite set of explicit founding beliefs or principle of physics.

He suggested that science was inherently amenable to this treatment since it held an internal recursion. Science is a description (of the universe) which exists within the universe. Thus a complete science must describe itself describing itself. Furthermore the universe imposes limits on the packing and storage of information. Hence Hawking concluded that physics should remain for ever incomplete. It is a nice sentiment but sadly the argument does not hold water. There are plenty of ways to escape these conclusions. For instance the no-input condition is equivalent to an infinite number of explicit axioms. Furthermore the Gödel arguments use classical logic. However, we have shown that the universe's natural logic happily accepts two indeterminate conclusions. Anyway the ultimate explanation of everything lies outside the mathematical bounds of science. It is a shock to the scientific mind but there is a non-quantitative foundation to science and reality.

So what is new in the hypothesis of eventual reality? There is no straightforward answer. To emphasize the point, let me start by being a little facetious. Every letter of every word that I have used is written everyday by most of us. Each word and even every pair of words have been used together too many times to count. Yet surprisingly most sentences will never have been uttered or written in the precise form given here. However, the sentiment these sentences communicate will have been expressed by innumerable others. Everyone can effortlessly produce much speculation. Someone somewhere (probably including an ancient philosopher in Egypt, India, China or Greece) will have previously proffered every idea as an option. Thus it is not the individual ideas that are special. It is how these ideas are assembled and where they lead that is new. Anaximander deduced many of the properties of the apeiron. Leibnitz recognized the apeiron's network-like form. Indubitably many philosophers and scientists of more recent eras have discussed similar structures. Certainly the networked apeiron is a critical element of the picture of eventual reality – but it is not what distinguishes the theory from previous speculation. Furthermore all the analytical methods I employed have also been used before. Indeed prior experience with these established analytical procedures was extremely beneficial to me during the development of the hypothesis. This familiarization is why it is easier for us than for earlier generations to recognize and appreciate eventual reality.

Therefore can we answer what marks eventual reality apart? Well there are many things. For instance eventual reality is founded on a tautology. It is a comprehensive theory of absolutely everything. It has a fully rational justification (presented here) and actually an equally impressive empirical justification (presented elsewhere). It incorporates the realm of science but extends beyond into metaphysics. It describes all the storeys of the edifice of reality.

Now, to understand the new picture, look at the previously accepted facts, laws, principles and beliefs that the hypothesis has demoted or displaced. In many ways these rearrangements characterize the hypothesis. Hence this self-justified organization of accepted ideas and their interplay defines the big picture given by eventual reality.

Then again, perhaps what is special about eventual reality is that it has no genuine rivals. Every other proposal, at a minimum, takes logic for granted. Most take mathematics for granted. Admittedly if we make reasonable presumptions, in developing a theory, we are likely to end up with something that looks reasonable. However, we should not do this when formulating fundamental theories. We cannot rely upon our experience of the everyday world to indicate reasonableness. Recall that the ancient Greeks thought space must be Euclidean. Today, many believe that the universe is the totality of space and its contents. Such presumptions once appeared obvious and without need of justification. However, any fundamental theory based upon such presumptions is speculative.

Within a scientific framework fundamental speculation inevitably assumes rules of logic, rules of mathematics, the existence of laws of nature, the existence of space (and time) and normally some quantitative entities or phenomena.

Philosophical speculation is equally presumptive. It invariably relies upon the existence of logic, mathematics, time and rationality. Compositionally the speculation is likely to involve material/quantitative monism, mental/semantic monism or neutral monism with unexplained correspondences.

Thus eventual reality may be a hypothesis currently without direct experimental supporting evidence, but eventual reality is not speculative – whereas all its supposed peers are. It is clear that, at best, these competitors are seeking to place fundamental representations on elevated storeys of reality. Eventual reality is not pie in the sky; after all, it starts at ground level.

FINALE

Ten thousand years ago an ancestor of mine, actually one of your ancestors too, was sitting around the campfire with his extended family group. The stars glistened in the sky, the wood crackled in the fire. The light created dancing shadows behind them. Our ancestor told a tale that his father (and his father's father before him) had handed down to him. He told a tale about the meaning of life, the origin of the universe and the struggle for supernatural supremacy. He told a tale rather like the one I have described here. Of course our ancestor had to change a few names to protect the innocent ears of younger listeners. He had to dramatize and personify some features to make the story more entertaining and mysterious. He had to invent a few bits of magic to make the recitation scarier. He had to leave out some of the more complicated details so that everyone could comprehend the delivered narrative.

Most of this epic legend has been lost. However, fragments remain scattered through out the world's recorded mythology. The multistorey edifice of reality became a tower of turtles. The profound tautological loop became a tail-devouring serpent, Uroborus. The first principle became a benevolent power. Today science is the dispenser of the wisdom of ages but science is uncomfortable with such story telling. Science prefers more precision and less drama. These attributes have created a science common to all societies. However, the resulting messages of science are less widely understood.

This ignorance of the workings of science can make the often-clinical scientific explanations seem heartless. It is our want as scientists to necessarily attempt to strip away any subjective viewpoints from our arguments to reveal objective truths. Some of these truths are quite stark. The humanist may seize upon an uncomfortable principle (say the second law of thermodynamics) and decry science for the senselessness that the humanist beholds. Note that the second law is quite unforgiving. This law states that the disorderliness of the material universe will increase indefinitely. Thus it may seem that everything in the universe is futile; it may seem that, in a billion aeons, everything will dissipate. Yet this inevitable decay need never occur. The reality might scare you even more. Decay only occurs in senseless situations. Thus our intelligence bestows a responsibility upon us. We will get to choose our future.

Furthermore science, the messenger, is far removed from senselessness or other clinical features of its messages. The pure scientists concentrate on discovering new understandings thereby enabling the mastery of mystery. Hidden behind this scientific dispassionate front is the genuine joy of observing the wondrous workings of the world. All can share this joy. What is more, every coin has two sides. After all, the second law is only a rewording of the first principle (for inanimate matter). The first principle delivers space, matter and experience. Thus a few applications of the first principle turn the turmoil of the apeiron into a marvellous world possessive of meaning, belief and feeling. How comforting is that?

Let us summarize some of the meanings we have found. We started with the definition that the universe is everything. This exclusivity implies that there are no external constraints placed upon the universe. Nothing makes the universes take any particular form; nothing initiates the universe nor prevents it from existing. Thus the fundamental universe is uninhibited opportunity. We found a way of interpreting this prospect as the apeironic network of monadic events. On this level, rampant opportunity reigns supreme.

This may seem like a recipe for disaster. However, we find orderliness in the statistics of this free-for-all. Procedurally we use the first-principle tautology. Regularity increases with every application of this principle. Each application builds another storey to reality. The first application of the principle produces space and aether. A second application produces matter and quantum mechanics. We mould information out of this increasing contextual certainty. There is one natural symmetry structure. Furthermore the symbols of the universal symmetry offer a natural language imprinted into the aether. Thus we discover that consciousness is the conflation of this apeironic opportunity with an elevated-storey certainty. We really are both the children and the masters of the universe. It is not only that the chemical elements (apart from hydrogen) building our bodies are the ash ejected from the embers of dying stars. Our awareness and decisions operate through the langue of the universe written upon the opportunist aethereal fabric. More than anything else all intelligent life represents the will of the universe.

GLOSSARY

algebra *(n.)* organized mathematical system that uses symbols to describe operations and elements.

apeiron *(n.)* an infinite non-spatial network of events without scale or direction. The apeiron is the foundation of reality.

anticommute *(v.)* have the property of reversing the resultant direction/sense when exchanged.

Aristotelian logic *or* **classical logic** *(n.)* a simple yes/no logic formally based upon three premises and described by Aristotle *cf.* logic.

boson *(n.)* any quantum *not* obeying the Pauli exclusion principle. These quanta have integral spin, *cf.* fermion.

classical *(adj.)* **1.** of or relating to classical mechanics. **2.** founded upon the beliefs of classical mechanics. **3.** of or relating to the logic of Aristotle.

classical mechanics *(n.)* a quantitative theory of a definite reality developed from Newton's picture of a universe comprising space, time, matter and motion.

commutative *(adj.)* having the property that expression ordering is unimportant.

commute *(v.)* have algebraic property of being able to be exchanged without consequence.

corpuscle *(n.)* minute material body.

cosmos *(n.)* the material universe comprising matter, space and motion *cf.* universe.

edifice of reality *(n.)* a conceptual multi-storey hierarchical sequence of expansive beliefs. Each storey describes a rational reality on a substantially different scale.

elementary particle *(n.)* one of those particles believed to be the basic constituents of all matter and/or the cause of all interactions.

event *(n.)* **1.** a monad that may only refer to other monads. **2.** something that happens in space-time.

eventual reality *(n.)* **1.** name of a hypothesis describing a universe based upon a foundation apeiron, and from which all of science can be derived. **2.** an ultimate foundation reality.

Fermat's principle *(n.)* the summing of microscopic contributions mostly produces cancellation or low-level noise, leaving the overabundant constructive sets to form the macroscopic picture.

fermion *(n.)* any quantum obeying Pauli exclusion principle. These quanta have half-integral spin, *cf.* boson.

Grassmann algebra *(n.)* a vector-like algebra with anticommuting elements.

logic *(n.)* any system of rules, constructed from Boolean components, that sentient creatures may appear to use in decision making.

macroscopic *(adj.)* relating to a reformulated larger-scale description ignoring some smaller-scale features *cf.* microscopic.

metaphysics *(n.)* the academic discipline that describes the non-quantitative nature of the universe *cf.* science and natural philosophy.

microscopic *(adj.)* relating to a detailed smaller-scale description *cf.* macroscopic.

monad *(n.)* an irreducible fundamental metaphysical part of the foundation of reality. There is only one type of monad.

monadic *(adj.)* of or relating to one or more monads.

natural philosophy *(n.)* the academic discipline that describes the nature of the universe *cf.* science.

Pauli exclusion principle *(n.)* statement that two identical quanta cannot have the same attributes eg occupy the same state. This principle applies to fermions.

quadron *(n.)* an umbrella concept containing precisely four distinct ordered components eg science is a quadron.

quadronic *(adj.)* of or relating to one or more quadrons.

quantal *or* **quantum** *(adj.)* **1.** of or relating to one or more quanta. **2.** quantum mechanical.

quantum *(n.)* **quanta** *(pl.)* minimum unit of particular entity by which system can be changed.

quantum mechanical *(adj.)* relating to the theory of quantum mechanics.

quantum mechanics *(n.)* quantitative scientific theory of dynamics and measurement applicable to systems of quanta eg elementary particles.

reality *(n.)* that which is believed to exist in a grand system of beliefs eg a worldview given by one storey of the edifice of reality. Scientific worldviews are rational; cultural beliefs often contain many irrational elements.

science *(n.)* the academic discipline that describes the quantitative nature of the universe *cf.* metaphysics and natural philosophy.

spin *(n.)* intrinsic angular momentum of a particle. Spin is quantized.

universe *(n.)* the totality of whatever is called real on *every* storey of the edifice of reality *cf.* cosmos.

Index

D

E

F

G

H

I

J

K

L

M

N

O

P

Q

R

S

T

U

V

W

Z

Made in United States
Troutdale, OR
11/03/2026

56844691R00100